Adaptive Constrained Control for High-Order Fully Actuated Nonlinear Systems

Changchun Hua • Liuliu Zhang • Pengju Ning

Adaptive Constrained Control for High-Order Fully Actuated Nonlinear Systems

 Springer

Changchun Hua ⓘ
Automation department
Yanshan University
Qinhuangdao, Hebei, China

Liuliu Zhang ⓘ
Automation department
Yanshan University
Qinhuangdao, Hebei, China

Pengju Ning ⓘ
Automation department
Yanshan University
Qinhuangdao, Hebei, China

ISBN 978-981-95-0961-4   ISBN 978-981-95-0962-1   (eBook)
https://doi.org/10.1007/978-981-95-0962-1

This work was supported by Engineering Research Center of Ministry of Education (633000404) and National Natural Science Foundation of China (U20A20187).

This Springer imprint is published by the registered company Springer Nature Singapore Pte Ltd.
The registered company address is: 152 Beach Road, #21-01/04 Gateway East, Singapore 189721, Singapore

If disposing of this product, please recycle the paper.

Preface

In the vast realm of modern control theory, the study of nonlinear systems has always held a pivotal role. With the rapid advancement in industrial automation, aerospace, robotics, and intelligent transportation systems, the demand for precise control of high-order fully actuated nonlinear systems is growing increasingly. These systems, characterized by their complex dynamic behaviors and high degree of nonlinearity, challenge the limits of traditional control strategies while also offering new opportunities and challenges for the development of control theory.

This monograph aims to delve deeply into the issue of adaptive constraint control for high-order fully actuated nonlinear systems, an area of study that combines both theoretical depth and broad application. We will begin with the fundamental theories of high-order fully actuated nonlinear systems, gradually progressing to the design of adaptive control strategies, and then to the implementation of constraint control, striving to provide readers with a comprehensive and systematic perspective.

In this book, we will first introduce the basic characteristics and mathematical models of high-order fully actuated nonlinear systems, laying a solid foundation for the subsequent design of control strategies. Subsequently, we will discuss in detail the fundamental principles of adaptive control theory, including parameter estimation, model reference, and adaptive laws, which are key technologies for achieving effective control. Particularly, we will focus on how to ensure system stability under input constraints and time constraints while achieving constraint control of system outputs or states, which is crucial for ensuring the safety and reliability of the system.

This book will provide a detailed analysis of the impact of these constraints on control system design and propose a series of innovative adaptive control strategies to meet these constraint conditions. We will discuss how to utilize advanced mathematical tools and computational methods, such as adaptive control theory, neural network technology, etc., to design control algorithms that can respond in real-time and adapt to dynamic changes in the system.

In addition to the above research, in the last chapter of this book, we will explore the practical applications of adaptive constraint control for high-order fully actuated nonlinear systems in real-world engineering scenarios, including but not limited to unmanned underwater vehicle system. By analyzing specific case studies, we will demonstrate how theoretical concepts can be transformed into powerful tools for solving practical problems.

This book is intended for researchers in control theory, engineers, and students majoring in related fields at universities. We hope that this book will not only provide readers with valuable theoretical knowledge and practical experience but also inspire further research and innovation in the control of high-order fully actuated nonlinear systems.

Chapter 1 introduces the basic concepts of high-order fully actuated nonlinear systems and provides a brief overview of adaptive control and constraint control. Chapter 2 introduces some essential mathematical preliminaries.

Then, the rest of this book will be presented under the following parts:

Part I: The first part of this book is concerned with high-order fully actuated nonlinear systems with input-constrained. In Chap. 3, the issue of time delay under high-order fully actuated nonlinear systems with uncertain parameters is investigated. In Chap. 4, the results are extended to dead zone constrained and unmodeled dynamics.

Part II: The second part of this book focuses on high-order fully actuated nonlinear systems with time-constrained. In Chap. 5, time-constrained control for strong interconnected high-order fully actuated nonlinear systems is investigated. In Chap. 6, adaptive time-constrained control for high-order fully actuated nonlinear systems is studied.

Part III: The third part of this book processes adaptive state-constrained control for high-order fully actuated nonlinear systems. In Chap. 7, event-triggered control for high-order fully actuated system under output constraint is investigated. In Chap. 8, the results are extended to the case of multi-variate constraints. In Chap. 9, adaptive tracking error-constraint control of fully actuated unmanned underwater vehicle systems is studied.

Qinhuangdao, China   Changchun Hua
May 2025   Liuliu Zhang
                                                                                      Pengju Ning

Competing Interests The authors have no competing interests to declare that are relevant to the content of this manuscript.

Contents

Notation

Notations Related to Subspaces

- $\mathbb{R}$ is the set of all real numbers.
- $\mathbb{R}^+$ denotes the set of all positive real numbers.
- $\mathbb{N}^+$ stands for the set of all positive integers.
- $\mathbb{R}^n$ is the set of all real vectors of dimension n.
- $\mathbb{R}^{m \times n}$ stands for the set of all real matrices of dimension $m \times n$.
- $\mathbb{C}^i$ denotes that the derivative of the i-th order exists and is continuous.
- $x^{(i)}$ is that the i-th derivative of the variable x.
- $|A|$ denotes the absolute value of scalar A.

Notations Related to Vectors and Matrices

- 0_n is the zero vector in $\mathbb{R}^n$.
- $0_{m \times n}$ denotes the zero matrix in $\mathbb{R}^{m \times n}$.
- 0 stands for the zero matrix of appropriate dimension.
- I_n denotes the identity matrix of order n.
- I is the identity matrix of appropriate dimension.
- A^{-1} stands for the inverse matrix of matrix A.
- $x^{(i)}$ is that the i-th derivative of the variable x.
- A^T is the transpose of matrix A.
- $\mathrm{Re}(A)$ denotes the real part of matrix A.
- $\lambda(A)$ stands for the set of all eigenvalues of matrix A.
- $\lambda_i(A)$ is the i-th eigenvalue of matrix A.
- $\lambda_{\max}(A)$ stands for the maximum eigenvalue of matrix A.
- $\lambda_{\min}(A)$ denotes the minimum eigenvalue of matrix A.
- $\mathrm{rank}(A)$ denotes the rank of $A \in \mathbb{R}^{m \times n}$.
- $\mathrm{tr}(A)$ is the trace of $A \in \mathbb{R}^{m \times m}$.
- $\|A\|_2$ is the spectral norm of matrix A.
- $\|A\|_F$ denotes the Frobenius norm of matrix A.
- $\inf(A)$ means the infimum of A.

For a vector $x \in \mathbb{R}^m$, a symmetric positive definite matrix P, a set of constants $A_i \in \mathbb{R}^{m \times m}, i = 0, \ldots, n-1, L = \begin{bmatrix} 0 \cdots I_r \end{bmatrix}^T$, we have the following definitions:

- $P_L = PL$.
- $\|x\|_P = \left(x^T P x\right)^{\frac{1}{2}}, \|x\| = \left(x^T x\right).$
- $x^{(0 \sim n-1)} = \begin{bmatrix} x, \dot{x}, \ldots, x^{(n-1)} \end{bmatrix}^T, \quad x_{i \sim j}^{(0 \sim n-1)} = \begin{bmatrix} x_i^{(0 \sim n-1)}, x_{i+1}^{(0 \sim n-1)}, \ldots, \\ x_j^{(0 \sim n-1)} \end{bmatrix}^T, \; j \geq i.$

- $A^{0 \sim n-1} = [A_0 \; A_1 \; \ldots \; A_{n-1}], \Phi(A^{0 \sim n-1}) = \begin{bmatrix} 0 & I & \\ & & \ddots \\ -A_0 & -A_1 & \cdots & -A_{n-1} \end{bmatrix}.$

Chapter 1
Introduction

1.1 Background

In the mid-seventeenth century, Swiss mathematician Leonhard Euler proposed the "order elevation and reduction" methodology when solving higher-order nonhomogeneous ordinary differential equations, which subsequently became regarded as the canonical approach for solving such equations. Over two centuries later, Hungarian-American mathematician Rudolf Emil Kálmán pioneered the state-space model, representing systems in the temporal domain through state variables and state-space formulations, while establishing fundamental theories of controllability and observability. This established the state-space model as one of the predominant paradigms for system representation. While this modeling framework proves advantageous for state response analysis, filtering, and observer design, it exhibits inherent limitations in control input computation and, more significantly, compromises the system's intrinsic fully-actuated properties. Fully-actuated systems (FASs) occur naturally in physical systems, yet their prevalence has been overshadowed by the more widespread underactuated systems, resulting in limited research attention. The control of FASs offers remarkable simplicity: their principal advantage lies in the direct elimination of all undesirable open-loop dynamic characteristics while simultaneously introducing new closed-loop dynamics that precisely meet design specifications with near-ideal performance. To better address system control challenges, Duan recently introduced the High-Order Fully-Actuated (HOFA) system model specifically tailored for controller design, establishing a unified framework for HOFA-based controller synthesis.

Although state-space models and HOFA system models can be mutually transformed, they represent two fundamentally distinct approaches to system description. The primary reasons are as follows: First, since fundamental physical laws such as Newton's laws, momentum (moment) theorems, Lagrange equations, and Kirchhoff's laws typically establish real-world systems as second-order FASs, different modeling approaches yield different representations: Applying the "increase vari-

© The Author(s) 2026
C. Hua et al., *Adaptive Constrained Control for High-Order Fully Actuated Nonlinear Systems*, https://doi.org/10.1007/978-981-95-0962-1_1

ables and dimensions while reducing order" principle leads to first-order state-space models. Employing the "eliminate variables and reduce dimensions while increasing order" method results in HOFA system models. Second, the state-space approach heavily relies on system-specific structural properties when solving controller design problems for nonlinear systems, making it difficult to obtain globally applicable results. In contrast, the HOFA method establishes a unified control system design framework that can conveniently design deterministic, performance-satisfying controllers for: linear systems or nonlinear systems, fully actuated or underactuated systems. Moreover, it ensures a globally consistent, time-invariant closed-loop system with desired eigenstructure. Third, for practical system models, the HOFA-based controller design procedure is remarkably straightforward, requiring only two steps: HOFA model (or sub-fully-actuated model). Design the corresponding HOFA controller (or sub-fully-actuated controller).

For more than half a century, the state-space approach has dominated the world of systems and control. Although the state-space method is more suitable for the derivation and observation (estimation) of state-response solutions, it is not as convenient as expected for the control problem. HOFA system theory is a new theoretical system for control system analysis and synthesis creatively proposed by Academician Duan Guangren. The main advantage of this method is that it can effectively eliminate most of the dynamic characteristics in the open-loop system that are not conducive to system analysis or control, including but not limited to linear and nonlinear characteristics, and in the process construct a new closed-loop dynamic system that can be flexibly configured for the position of the pole of the system. The theoretical system is based on the HOFA system model of the control system to achieve analysis and synthesis. The HOFA system model is a general control system model, which can be obtained directly by modeling the actual system by using the basic physical laws such as the Lagrange equation, Kirchhof's law, momentum and angular momentum theorem, and Navier-Stokes equation, or indirectly by performing differential homemorphism transformation or elimination upgrade transformation of under-flooding systems such as strict feedback systems and affine systems. At the same time, as a universal method paralleled to the state-space method, the research content of HOFA system theory involves many aspects such as system modeling, analysis, design and application, and covers all levels of control disciplines. An important advantage of the HOFA method is that a linear and stationary closed-loop system with the desired characteristic structure can be obtained through state feedback, and the parametric method also provides all the degrees of freedom in this design, once the HOFA system model of the dynamic system is derived, the HOFA characteristics of the system can be used to eliminate the nonlinear terms in the system, so as to obtain a constant linear closed-loop system. The proposed HOFA system approach does not require the conversion of a high-order system to a first-order system, and for a particular system design, the HOFA system approach requires fewer steps than the usual first-order system approach, and is therefore generally more straightforward and simple. Compared with the traditional state-space model, the HOFA system model does not need to be downgraded, which avoids the shortcomings of the original system such as loss of physical meaning, pathological matrix, and instability of numerical solutions caused by model downgrading, and also maintains the important properties for system analysis, such as controllability and stability.

The core control challenge resides in determining control inputs rather than state estimation. The inherent full actuation ensures analytical solvability of control variables, significantly simplifying controller design and enabling linear time-invariant closed-loop systems with global exponential stabilization to be realized. Although physically perfect FASs remain relatively scarce in practice (with many systems being underactuated), recent theoretical advancements have mathematically generalized the concept of full actuation. Through persistent research efforts, many underactuated systems can now be transformed into HOFA representations, thereby substantially expanding the application scope of HOFA system theory.

In recent years, scholars have developed HOFA controllers for various non-linear systems. In reference Duan (2021b), Duan proposed higher-order HOFA strict-feedback systems by extending traditional first-order architectures, establishing corresponding HOFA models and controllers. Based on rank conditions, the pseudo-linear system was then converted into a canonical form, which was further transformed into a HOFA model. For systems with uncertainties, researchers have applied the HOFA approach to controller design and synthesis. Without requiring prior knowledge of the maximum or minimum eigenvalues of the unknown control input matrix, they designed an HOFA adaptive event-triggered controller that ensures the boundedness of all closed-loop system signals while avoiding Zeno behavior. Subsequently, a HOFA controller was designed to endow the closed-loop system with the desired eigenstructure. Cai et al. (2023c) presented a novel observer-based fault-tolerant controller framework for a class of nonideal time-varying HOFA system. By skillfully fusing the techniques of concurrent learning and HOFA approach, Liu et al. (2023b) proposed a novel controller. A HOFA model was introduced to solve the coordinative control problem of networked nonlinear multi-agents in Liu (2022b). Meng et al. (2022a) investigated the adaptive event-triggered control problem for a class of uncertain HOFA systems. In practical systems, the HOFA controller has been successfully applied in various fields, for example, DC microgrids (Yu et al. 2024), multirotor aerial vehicles (Lu et al. 2023; Ricardo Jr & Santos 2022).

1.2 Fully Actuated Systems

In the history of control theory research, the state-space approach established by Hungarian-American mathematician Rudolf Emil Kalman has long held an overwhelmingly dominant position. By introducing first-order state-space models into system analysis and control research, this methodology naturally places the system's state variables at the core of investigation while relegating control variables to a secondary role. Within this framework, problems such as state variable solution, response analysis, and related estimation can be readily addressed. However, the convenience of solving control variables and their physical realizability have been largely overlooked. To facilitate straightforward derivation of analytical solutions for control variables, Academician Guang-Ren Duan systematically found the HOFA system theory. He comprehensively elaborated on the theory's physical background, fundamental definitions, theoretical framework, and application scope. Furthermore, he reviewed and resolved classical problems in control theory research under this new paradigm, including robust control, disturbance rejection and decoupling, optimal control, and time-delay system control.

The HOFA system approach, has demonstrated significant advantages in addressing control problems of dynamic systems, particularly for strict-feedback nonlinear systems. Distinct from conventional methods requiring system order reduction, the HOFA approach enables direct controller design for higher-order systems while retaining their inherent structural characteristics. This methodology achieves system linearization through pseudo-linear state feedback control laws, transforming closed-loop systems into linear time-invariant systems with specific configurations. Within this framework, controller design can be systematically implemented via parameterization methods by configuring coefficient matrices of the characteristic polynomial to achieve desired system specifications. This approach not only streamlines the controller design process but also enhances system controllability and performance. Current research on HOFA system methodology primarily focuses on five key aspects:

1. Modeling of FAS: The fully-actuated property, characterized by direct one-to-one correspondence between control inputs and state variables, enables significantly greater design flexibility in control system development. The dynamics of most engineering systems originate from fundamental physical laws such as Newtonian mechanics, Lagrangian dynamics, momentum conservation principles, and Kirchhoff's circuit laws. Modeling based on these first principles naturally yields second-order Lagrangian systems or higher-order fully-actuated representations, preserving the intrinsic system structure without approximation. In Zhao and Duan (2022), a six-degree-of-freedom FAS model for rigid spacecraft was established through variable elimination method to address position and attitude tracking control problems. Duan (2021b) generalized strict feedback systems containing parameter vectors and state vector derivatives as generalized strict feedback systems, proposing a recursive method to equivalently convert high-order generalized strict feedback systems into HOFA system models. However, there exist numerous special nonlinear systems in current engineering applications, such as underactuated systems, upper triangular systems, stochastic systems, and fractional-order systems. Converting these special systems into FAS models constitutes the foundation for widely applying HOFA system methodologies.

2. Parametric Design Method: Parametric design is an innovative control strategy for HOFA. By appropriately selecting and designing controller parameters, it enables precise adjustment of system dynamic performance, such as response speed, stability margin, disturbance rejection capability, and adaptability to

uncertainties. Duan (2021c) investigated the robust control problem for uncertain high-order nonlinear systems, proposing a HOFA model-based approach to design robust stabilizing controllers and robust tracking controllers. Duan (2021d) introduced direct design methods for adaptive stabilizing controllers and adaptive tracking controllers, which eliminate the need for converting high-order systems into first-order systems, demonstrating more direct and effective solutions. Building on these studies, Duan (2021e) addressed HOFA systems with nonlinear uncertainties and time-varying unknown parameters, developing a robust adaptive control method based on Lyapunov stability theory.

3. Research on Controllability and Observability of High-Order Systems: For HOFA, a comprehensive analysis of controllability and observability must be conducted to determine whether the system possesses full actuated characteristics. This analysis is then leveraged to optimize the system's control structure. Based on the unique characteristics of HOFA models, Duan (2021g) proposed the concept of controllability for general dynamical control systems, further revealing that a general dynamical control system may consist of a controllable subsystem described by a HOFA model and an additional uncontrollable subsystem.

4. Controller Design Based on Lyapunov Stability Theory: For nonlinear HOFA systems derived from physical law modeling or state-space model transformation, it is essential to construct suitable Lyapunov functions and corresponding intermediate control laws to ensure global stability of the closed-loop system. Duan (2022b) transformed the optimal control problem into a linear quadratic regulation (LQR) problem. By designing a composite controller, nonlinear issues were simplified into linear ones, thereby enabling the derivation of optimal control laws via the Riccati equation. In the context of infinite-time output regulation, the feasibility of HOFA control was guaranteed by ensuring that system initial conditions met specific requirements. Duan (2022a) introduced fundamental theories for discrete-time HOFA systems, proposing both step-forward and step-backward types of HOFA models, and designed state feedback controllers for these two model categories. When discussing the design of generalized PID control and tracking controllers, Duan (2022c) emphasized the critical role of constructing Lyapunov functions. It rigorously proved that the closed-loop system maintains stability and achieves robust tracking performance despite external disturbances and time-varying system parameters.

5. Engineering applications: The HOFA system methodology is employed to address practical engineering challenges, such as spacecraft attitude control and multi-robot cooperative control, demonstrating its versatility in real-world implementations. Expansion and optimization of HOFA system methods in complex engineering applications requires not only theoretical feasibility considerations, but also meticulous control strategy adaptation and optimization based on specific engineering contexts and practical operating environments. This involves achieving rapid response and precise tracking under constrained resource conditions, while enhancing control performance amidst complexities

such as noise, uncertainties, and model mismatches to ensure system reliability and stability.

The core advantage of the HOFA method lies in its ability to fully preserve the high-order characteristics of systems while applying linear system theory for controller design. This approach provides greater design freedom and flexibility, demonstrating particularly outstanding performance when dealing with systems exhibiting complex dynamic behaviors and pronounced nonlinear features. The HOFA methodology enables deeper understanding and control of system behaviors by preserving higher-order attributes, effectively overcoming limitations inherent in conventional approaches that focus solely on lower-order characteristics, such as oversimplification or neglect of critical dynamic details. Consequently, the HOFA method demonstrates remarkable technical advantages and broad applicability, both in theoretical analysis and practical implementation.

Currently, the HOFA methodology has received considerable research attention from scholars worldwide. Lu et al. (2023) introduced an enhanced HOFA framework designed for multiple high-order dynamic systems such as quadrotor platforms, eliminating the dependence on pseudo strict-feedback structures that were essential in conventional HOFA methodologies. Adıgüzel and Yalçın (2022) investigated disturbance suppression in discrete-time control architectures for fully actuated mechanical systems, specifically focusing on compensation strategies for input perturbations. Hu et al. (2023) developed a parametric time-variant canonical interval observer by synergistically integrating linear time-varying transformation methodologies with closed-form solutions of nonstationary fully-actuated Sylvester differential equations. Liu et al. (2023b) presented an innovative concurrent learning-enhanced hybrid control architecture through synergistic integration of predictive concurrent learning mechanisms, adaptive nonlinear command filtering techniques, and HOFA paradigms, ultimately synthesizing a multiscale controller with closed-loop stability guarantees and dynamic performance optimization capabilities. Hu et al. (2023) proposed a novel adaptive guaranteed-cost tracking controller for HOFA systems with time-varying uncertainties. Tian et al. (2024) proposed a composite controller grounded in a FAS framework to address the challenge that system uncertainties and diverse physical constraints severely degrade trajectory tracking performance. The controller integrated a nonlinear disturbance observer within the inner loop and a high-precision trajectory controller in the outer loop. Li and Duan (2023) proposed a HOFA control strategy for flexible servo systems by leveraging the singular perturbation method. To address the terminal impact-angle constraint in intercepting highly maneuvering targets, Chen et al. (2023) developed a three-dimensional guidance law by integrating a HOFA framework with a cascaded linear extended state observer architecture. Cui et al. (2024) presented a discrete-time HOFA model reference tracking control framework for combined spacecraft simulators operating under external disturbances.

In the context of adaptive control, Zhang et al. (2023e) addressed the control problem of nonlinear HOFA systems with time delays, unmodeled dynamics, and unknown dead-zone input characteristics. Zhao et al. (2023c) investigated adaptive control of nonlinear HOFA systems with nonstrict-feedback structure, involving both uncertain constant and time-varying delays. Wang et al. (2023b) proposed an adaptive stabilization control scheme for uncertain high-order fully actuated nonlinear systems under disturbances. Ning et al. (2022b) addressed the adaptive control problem for a class of time-delay systems by developing an adaptive controller grounded in the fully actuated system approach, achieving asymptotic stability of the closed-loop system. Liu et al. (2023a) investigated the adaptive control problem for hybrid HOFA systems with unknown parameters. By employing tuning functions within the HOFA system framework, an adaptive control law was developed that eliminates the need for overestimation of unknown parameters, thus guaranteeing closed-loop stability. Liu et al. (2023c) developed a HOFA command-filtered adaptive controller within the fully actuated system framework. This controller was applied to track reference signals for both second-order and higher-order strict-feedback systems with parametric uncertainties. In contrast to conventional backstepping methods, the proposed approach effectively circumvents the differential explosion phenomenon through its command-filtered architecture.

In the domain of fault-tolerant control, Liu et al. (2022e) proposed an adaptive fault-tolerant control strategy for a class of HOFA systems. This methodology introduced an observer to adaptively estimate system states and actuator faults, while rigorously analyzing the impact of observation errors on stability through singular value perturbation analysis of system matrices. Cai et al. (2023a) developed an active fault-tolerant control framework for HOFA systems with uncertainties. By establishing a HOFA adaptive observer-controller architecture, this work derived a closed-loop HOFA dynamic data model structure. Furthermore, an active fault-tolerant control strategy was proposed for HOFA systems by exploiting the fault information embedded in the dynamic data model. Cai et al. (2023b) developed a fault-tolerant trajectory tracking control scheme for nonlinear HOFA systems under simultaneous multiplicative and additive actuator fault conditions. Dong et al. (2023) addressed the adaptive tracking control problem for uncertain HOFA systems subject to actuator faults and full-state constraints. By designing a novel nonlinear transformation function that depended solely on system states and constraint boundaries while handling asymmetric time-varying constraints, an equivalent unconstrained HOFA system model was derived. Building upon this transformed model, an adaptive fault-tolerant controller was further developed through the HOFA system methodology. Cui et al. (2023b) developed an adaptive fuzzy fault-tolerant tracking controller by integrating HOFA system theory, dynamic surface control techniques, and fuzzy logic system methodologies. Ma et al. (2023) investigated the formation control problem for heterogeneous multi-agent systems (MAS) affected by actuator faults. To prevent fault propagation across the agent network, a hierarchical design methodology was employed. In the upper-layer architecture, a distributed prescribed-time observer was developed to estimate the leader's trajectory over directed communication graphs. In the lower-layer design, an adaptive fault-tolerant controller based on FAS theory was constructed for the heterogeneous MAS with actuator failures. This methodology guaranteed prescribed-time convergence of formation errors to zero while enhancing the robustness of HOFA theory against parameter uncertainties, actuator faults, and measurement noises. Cai et al. (2023c) developed a nonlinear augmented state observer and a low-power fully actuated control algorithm for non-ideal time-varying HOFA systems.

In the domain of predictive control, Liu (2022a) developed a nonlinear predictive controller for discrete-time nonlinear systems with time-varying delays, featuring a compact and generic framework for delay compensation. Building upon this foundation, Liu (2022b) extended the framework to address multi-agent cooperative control problems under time-varying communication delays while maintaining the core delay compensation mechanism. Zhang et al. (2022a) investigated the cooperative control problem of higher-order fully actuated multi-agent systems. By introducing a Diophantine equation to construct an incremental HOFA prediction model, this work further employed a model predictive control methodology to devise an optimal cooperative controller for distributed multi-agent coordination. Zhang et al. (2023a) addressed the cooperative control problem for HOFA multi-agent systems with input saturation and prescribed performance constraints. An optimal constrained cooperative controller was developed to rigorously guarantee compliance with the specified system constraints. In Zhang et al. (2023d), the consensus control problem for HOFA networked multi-agent systems with communication delays between network nodes and sensors, as well as between network nodes and actuators, was studied. A proportional-integral-based predictive control approach was proposed. In this approach, a local PI feedback controller was used to stabilize the closed-loop system, and multi-step output prediction was applied for cost function optimization, leading to the development of an optimal consensus controller. In addressing the output tracking problem for discrete HOFA systems, literature Zhang et al. (2023b) proposed a HOFA predictive controller and implemented it in spacecraft simulator control systems. For networked HOFA systems affected by communication delays and external disturbances, Zhang and Liu (2023) developed a disturbance-observer-based fully actuated predictive controller to handle the output tracking challenge.

In spacecraft control applications, Liu et al. (2022a) addressed the optimal attitude control problem for rigid-body spacecraft subject to actuator saturation constraints. The methodology began by formulating a second-order FAS representation of the attitude error dynamics through Modified Rodrigues Parameters. To compensate for aggregated disturbances arising from inertial parameter uncertainties and exogenous perturbations, an Extended State Observer is employed for real-time disturbance estimation. Building upon this framework, a fully-actuated control scheme is subsequently developed to achieve optimized attitude tracking performance. Xiao and Chen (2022) addressed the attitude tracking control problem for spherical liquid-filled spacecraft. The study established rigid-fluid coupling dynamic equations based on the spacecraft's Euler angles and the angular velocities of the liquid fuel. Utilizing FAS methodologies, these dynamic equations were systematically transformed into third-order and second-order differential equations governing the Euler angles. Subsequently, a fully-actuated controller was developed to achieve precise attitude control objectives under this framework. Xiao and Chen (2023) developed a second-order quaternion system for rigid spacecraft attitude dynamics based on FAS theory. Building upon this formulation, two closed-loop control strategies were systematically proposed to achieve spacecraft tracking control objectives with enhanced stability guarantees. Zhao and Duan (2022) presented a systematic framework for spacecraft attitude control through fully-actuated system modeling. The methodology begins by employing state transformation and variable elimination techniques to derive a reduced-order FAS model from the original nonlinear dynamic equations governing spacecraft attitude dynamics. Building upon this model, a fully-actuated controller is strategically designed to achieve dual objectives: precise attitude stabilization and agile maneuvering control with enhanced dynamic decoupling.

Domestic/international advances in HOFA system theory demonstrate their critical role in guiding framework development, theoretical refinement, and engineering implementation. However, the existing literature still exhibits some shortcomings and areas requiring improvement in the research on HOFA system theory and its applications, which can be summarized in the following four aspects:

1. Current literature has indicated that typical nonlinear systems with specific structures, such as strict-feedback systems, affine systems, and feedback-linearizable systems, can be transformed into HOFA systems. However, for general nonlinear systems, practical conditions to determine whether they can be converted into HOFA systems are still lacking.

2. Most literature primarily focuses on the stabilization control problem based on HOFA system theory, while paying less attention to more practically significant issues such as tracking control and output regulation. Some studies on tracking control convert the problem into stabilizing the error system, which ultimately achieves the desired tracking control objective but neglects the regulation of dynamic processes.

3. As a control-oriented system model, the HOFA system's advantage lies in its ability to more easily obtain a complete analytical form of the control law. However, the majority of the literature fails to consider input constraints in control design, such as saturation, dead zones, and energy limitations, with only a small number of studies partially addressing these issues.

4. In the application of HOFA system theory, the engineering objects are primarily aerospace vehicles directly modeled as HOFA systems. There is a lack of application cases in typical underactuated/overactuated systems, such as underactuated crane systems or underactuated robotic systems. Moreover, among the existing applications, only a few studies have conducted experimental validation on hardware-in-the-loop simulation platforms, while the rest remain limited to numerical simulations.

1.3   Brief Overview of Adaptive Control

The primary concept of adaptive control is to automatically adjust the controller parameters to address system uncertainties and thereby achieve the desired closed-loop performance. It typically comprises two components: identification and feedback control. The identification component relies on real-time state measurements from system sensors to estimate uncertainties online, while the feedback control component applies control actions based on the information obtained from the online estimation as well as the current system state measurements (see Krstic et al. 1995). Owing to the powerful online estimation capability inherent in the identification process, adaptive control demonstrates significant advantages in handling uncertainties and external disturbances, and it has been widely applied in various fields including defense and military, process industries, and aerospace. Self-tuning control and model reference adaptive control are the primary representatives of adaptive control methodologies.

Self-tuning control is a method capable of automatically detecting deviations or anomalies during operation and correcting them. By employing built-in feedback mechanism and algorithm, it continuously monitors the system's state and output, and automatically adjusts control parameters or actions in accordance with preset target values or standards to ensure that the system remains at the desired state or performance level. In the 1950s, the renowned scholar Professor Kalman first introduced the concept of self-tuning control, which combined system identification with control. Peterka (1999) later extended this idea to linear discrete single-input single-output (SISO) systems with unknown constant parameters. Shortly thereafter, Åström and Wittenmark (1973) designed an easily implementable self-tuning regulator based on minimum variance theory. However, this approach was limited to specific uncertain nonlinear systems, featuring a singular structure and neglecting engineering constraints. Clarke and Gawthrop (1975) subsequently developed a generalized self-tuning controller to overcome the limitations of the earlier algorithm. Moreover, Edmunds has made notable contributions to the development of pole-assignment-based self-tuning control algorithms. Agarwal and Seborg (1987) presented a design methodology for self-tuning regulators applicable to SISO systems with unknown parameters. Over the past two decades, various self-tuning control algorithms tailored to different types of systems have been proposed and implemented, see Jiang and Jiang (2017) and the references therein. Model reference adaptive control (MRAC) transforms the adaptive control problem into one of model matching, enabling the controlled system's output (actual response) to approximate that of a predefined reference model (desired response). Through continuous adjustment of the controller's parameters, the actual output is brought ever closer to the reference model output, thereby achieving adaptive optimization of system performance in the face of uncertainty and changing environments. In an effort to improve the control performance of aircraft autopilots, the scholar Whitaker of the Massachusetts Institute of Technology ingeniously introduced a model reference adaptive design approach. However, because this method did not guarantee global asymptotic stability of the controlled system, Butchart, Shachcloth, Parks, and Phillipson later incorporated Lyapunov stability theory into the MRAC framework to address the challenges of global stabilization. Building on this advancement, Laudau, Monopoli et al. developed a new MRAC algorithm that leverages augmented error signals and Popovhyper stability theory to adjust control parameters solely based on the system's input-output information, thereby eliminating the need for differential signals. Subsequent refinements have attracted widespread attention to MRAC, which continues to evolve. Although both self-tuning control and MRAC methods have demonstrated remarkable performance for linear systems with uncertainties, their effectiveness diminishes when dealing with significant nonlinearities and strong uncertainties—issues that can lead even to oscillatory behavior. In recent years, the rapid development of intelligent control theories, such as neural networks and fuzzy control, has shown extraordinary promise in handling highly nonlinear and severely uncertain systems. Consequently, many researchers have integrated these approaches to propose novel adaptive control schemes, which have been applied to robotic arms, power systems, mobile robots, and other applications (see Li et al. 2016; Roy et al. 2016; Nguyen and Dankowicz 2015). While existing adaptive control methods have achieved many important results, challenges remain regarding the complexity of control structures and the necessity for parameters to meet linearization conditions, thereby highlighting the need for further research.

By incorporating mathematical tools—specifically, results from differential geometry and differential algebra—into nonlinear system control theory, Isidori et. al introduced the feedback linearization control method. This innovation not only accelerated the development of nonlinear control theory but also stimulated further advancements in adaptive control methods for uncertain nonlinear systems. Building on the feedback linearization approach, adaptive control has achieved several significant results (see Sastry and Isidori (1989); Teel et al. (1991); Sastry and Bodson (2011)). However, these achievements are effective only for uncertain nonlinear systems that satisfy the matching condition or its generalized forms. To overcome these limitations, Kanellakopoulos et al. (1991) innovatively proposed the backstepping method, which enabled global adaptive control for a class of strict feedback uncertain systems. This approach not only relaxed the requirement of a unitary relative degree inherent in passive design methods but also circumvented the restrictions imposed by the growth conditions of nonlinear terms and the matching constraints. Therefore laying a solid foundation for the development of adaptive control theory for strict feedback uncertain nonlinear systems. While the backstepping method simplifies controller design, it introduces the challenge of parameter overestimation. Specifically, if an n-th order system has m unknown parameters, nm parameter estimates are required, and this overparameterization issue becomes more pronounced as the order of the system increases. By introducing a tuning function, Krstić et al. (1992) reduced the number of adaptive estimation parameters to match the actual number of unknown system parameters, thereby effectively mitigating the overparameterization problem. Subsequently, through the collective efforts of many researchers, a strategy was developed that employed a single adaptive parameter to estimate the maximum of all unknown parameters, thus reducing the dynamic order of the adaptive control law to a minimum, see Yu and Lin (2016). Moreover, based on adaptive backstepping control techniques, a variety of control methods have been derived for strict feedback uncertain nonlinear systems, including adaptive sliding mode control (Plestan et al. 2010), adaptive neural network control (Liu et al. 2021d), and adaptive fuzzy control (Chen et al. 2019). In general, the above-mentioned results formulated through first-order differential equations, relying on the celebrated backstepping method, often implying a relatively complex controller design.

1.4   Brief Overview of Constrained Control

The stable, reliable, and safe operation of systems constitutes one of the primary objectives pursued in nonlinear system control (Perez-Arancibia et al. 2010). Due to the inherent physical limitations of controlled objects and their surrounding environments, as well as the subjective requirements imposed by designers, practical nonlinear systems inevitably encounter constraints arising from multiple factors. In this context, control problems governed by various constraint conditions are collectively termed constrained control problems. Nonlinear system constraints primarily encompass three categories: **input constraints**, **state constraints**, **time constraints**.

1.4.1   Brief Overview of Input Constraints Control

In control systems, actuators often exhibit input constraints such as **saturation constraints**, **dead-zone constraints**, **input delay constraints**, **hysteresis**, **intput quantization**, and **actuator failure constraints**. For instance: In electromechanical systems, input voltage is subject to saturation constraints imposed by power supply voltage limits. In hydraulic servo systems, dead-zone constraints arise from the nonlinear characteristics of hydraulic valves. The presence of these input constraints may degrade control performance or even destabilize the system. Consequently, input constraint handling has long been a focal point in nonlinear control research, yielding substantial theoretical and practical advancements.

For nonlinear systems subject to **saturation constraints**, analyzing and compensating for their effects on system performance constitutes a central challenge. To address input saturation, researchers have proposed a variety of methodologies. To date, significant efforts have been devoted to addressing input constraints, with a primary focus on transforming saturated inputs into designable nominal inputs through systematic mappings. In Cheng et al. (2019) and Wang and Yang (2018), adaptive estimation methods and Nussbaum functions were employed to handle unknown control directions, enabling adaptive prescribed performance control for nonlinear systems with both unknown control directions and input saturation. For fractional-order strict-feedback and non-strict-feedback nonlinear systems subject to saturated inputs, Song et al. (2021b) and Cao and Nie (2021) developed adaptive neural network (NN)-based control schemes. Building on nonlinear transformations and auxiliary systems, Wang et al. (2022b) proposed an adaptive asymptotic tracking control framework for strict-feedback nonlinear systems with simultaneous state constraints and input saturation. By integrating multiple Lyapunov-Krasovskii functionals, Nussbaum gain techniques, and NN approximation, Li et al. (2023b) addressed controller design challenges in switched systems caused by unknown time delays, actuator saturation, and uncertain nonlinearities. For fractional-order nonlinear systems with input saturation, Cui and Tong (2023) introduced an event-triggered predefined-time output feedback control scheme using dynamic surface control (DSC) technology. This approach ensured semi-global practical predefined-time stability and guaranteed boundedness of all signals within a predefined time interval. Current methodologies for handling input saturation constraints primarily fall into two categories: Constructing auxiliary systems to compensate for saturation effects (Wang & Yang 2018; Cao & Nie 2021; Wang et al. 2022b; Song et al. 2020); Approximating saturation nonlinearities via smooth functions (Cheng et al. 2019; Song et al. 2021b; Li et al. 2023b; Cui & Tong 2023).

A prevalent nonlinear constraint in control system actuators and acquisition devices is the **dead-zone constraint**. To address this challenge, researchers have developed two primary categories of control methodologies: The first category involves designing dead-zone inverse models to mitigate the adverse effects of dead-zone constraints. Representative approaches include: A novel tuning-function backstepping control scheme proposed in Su et al. (2017), which accommodates actuator faults/failures and dead-zone constraints via inverse deadzone modeling. Adaptive inverse compensation control for flexible single-link manipulators with unknown deadzones and actuator failures (Zhao et al. 2023e). A generalized dead-zone adaptive inverse compensator for asymmetric deadzone constraints (Zhang & Tong 2021). Adaptive fuzzy control for interconnected autonomous vehicle platoons with position deadzones and state constraints, combining deadzone inverse techniques and tangent barrier Lyapunov functions (Tan-BLFs) (Wei et al. 2023). Dynamic threshold control enhanced by dead-zone inverse and Nussbaum functions to improve resource utilization efficiency (Liu et al. 2024a). The second category simplifies dead-zone limitations through nonlinear function approximations: A pioneering adaptive switching dynamic surface control (DSC) strategy for fractional-order non-strict feedback nonlinear systems with unknown deadzones and arbitrary switching (Sui et al. 2021a). Decentralized output feedback control for fractional-order nonlinear large-scale non-strict feedback systems with unmeasurable states and unknown deadzones (Zhan et al. 2023). Finite-time command-filtered backstepping control for nonlinear systems with deadzones, achieved by transforming asymmetric dead-zone models into linear systems (Cai et al. 2024). Approximation of non-smooth input saturation and dead-zone nonlinearities via non-affine smooth functions, converted to affine forms using mean-value theorems (Kang et al. 2022; Zhang et al. 2023g).

The widely existing **input delay** can not be ignored in practical engineering, which will lead to a decrease in control performance and potentially system instability. In Bekiaris-Liberis and Krstic (2016), the predictor-based control design scheme was developed for controlled systems in presence of input delay. Several results assume exact model knowledge, as the use of predictor feedback is more amenable for systems with known dynamics (Krstic 2008; Lin & Fang 2007; Mazenc & Bliman 2006). Some results have also focused on designing robust and adaptive controllers for linear uncertain systems with known delay. In Wu et al. (2025), the Pade approximation method was utilized to create an intermediate variable to solve the input delay problems. However, there exist certain limitations in the Pade approximation method, which can only handle the constant and small input delay. Subsequently, the auxiliary system was introduced in Ma et al. (2020) to compensate for the influence of long time-varying input delay. It should be emphasized that the input delay is still required to be known. Mazenc et al. (2012) proposed methods to construct Lyapunov–Krasovskii functionals for linear time invariant systems with additive disturbances. A finite dimensional nonlinear periodic controller for the SISO linear time invariant system with uncertain gain and input time delay was proposed in Gaudette and Miller (2014). However, the results for uncertain linear systems do not extend trivially to uncertain nonlinear systems and hence, a major challenge lies in the use of predictor feedback for input delay nonlinear systems with parametric uncertainties, unmodeled dynamics, external disturbances and unknown delays. For a class of uncertain nonlinear systems with additive disturbance and unknown constant input delay, Jain and Bhasin (2015) proposed an adaptive control law, which included an adaptive feedforward term based on the desired trajectory to compensate for the parametric uncertainty in the plant and a robust delay compensation feedback term based on the integral of the past control values.

In practical engineering, **hysteresis** nonlinearity may affect the control performance of a system. Since the hysteresis nonlinearity is non-differentiable, it may cause inaccuracy or oscillations and even lead to instability. In order to address such a challenge, a lot of approaches have been proposed. The most common approach is to construct an inverse operator to compensate the effects of the hysteresis (Tao & Kokotovic 1995; Liu et al. 2014) and the references therein. Robust control is developed by combining the inverse compensation for a novel dynamic hysteresis model in magnetostrictive actuators in Liu and Su (1998). In addition, approaches without constructing an inverse model have also been developed in Wen and Zhou (2007). Recently, numerous scholars have incorporated the backlash-like (B-I) hysteresis factor into their investigations on stochastic nonlinear systems (Zhu et al. 2021), switched interconnected nonlinear systems (Zhao et al. 2022c) and nonstrict feedback systems (Liu et al. 2021c). Gao et al. (2014) proposed an adaptive control scheme based on the backstepping design for a class of stochastic nonlinear systems with unmodeled dynamics and time-varying state delays. Su et al. (2005) developed an adaptive variable structure control strategy for a class of continuous-time nonlinear dynamical systems with hysteretic nonlinearities characterized by the Prandtl-Ishlinskii (P-I) model. Similarly, in interconnected systems, the information exchange among the subsystems often induces time delay. Using the Backstepping method, Hua and Guan (2008) developed adaptive laws and output feedback controllers for nonlinear interconnected time-delay systems.

Quantized control is a methodology employed to represent and modulate control signals using discrete values, and has garnered significant interest across disciplines like engineering and automation. Quantized control offers the dual advantages of reducing communication rates while maintaining control stability, thus attracting substantial research attention in recent years (Zhao et al. 2023d). Parallel distributed compensation event-triggered dynamic output quantization control was considered for the discrete-time switched Takagi–Sugeno (T–S) fuzzy systems (Yang et al. 2022b). A synthesized control approach was introduced, incorporating input and output quantizations to address nonlinear systems with mismatched parametric uncertainties using backstepping design, as documented in Zhang et al. (2022c). A co-design strategy was put forward to simultaneously determine the event-triggered parameter and the quantization scheme (Zhao et al. 2022b).

Another critical constraint in practical applications is the **actuator fault constraints**. To enhance system safety and ensure stability under such limitations, the development of fault-tolerant control (FTC) schemes is imperative. Key contributions include: Wang and Yuan (2020) investigated integrated guidance and control for bank-to-turn missiles subject to rapidly varying actuator faults and coupled multi-source uncertainties. Tong et al. (2014) designed an observer-based fuzzy FTC method for integer-order nonlinear systems. Li and Tong (2017) proposed adaptive fuzzy FTC with output constraints, addressing both failure and stuck-type actuator faults. Yang et al. (2022a) developed an adaptive FTC scheme for Multiple-Input Single-Output (MISO) strict-feedback commensurate fractional-order nonlinear systems with finite actuator fault occurrences, whereas Li et al. (2021) considered scenarios with infinite fault occurrences. Practical controller design must simultaneously address actuator faults and input nonlinearities. Recent advances include: Wang et al. (2021) proposed an adaptive FTC framework for strict-feedback commensurate fractional-order systems under input dead-zones and infinite actuator faults. Moradvandi et al. (2020) introduced finite-time FTC for integer-order systems with five types of unknown input nonlinearities, infinite actuator faults, and arbitrary time-varying output constraints. Pishro et al. (2022) extended the methodology of Moradvandi et al. (2020) to incommensurate MISO fractional-order systems.

1.4.2 Brief Overview of Time Constraints Control

Convergence speed is a crucial performance metric in control systems, referring to the time required for a system to transition from its initial state to a stable or target state. In control theory, enhancing the convergence speed is often a key objective in controller design, particularly for applications demanding rapid response and high efficiency, such as robotic motion control, aircraft attitude control, power system stability control, and chemical process control (Galicki 2015; Na et al. 2015). While asymptotic control has achieved significant research progress in the aforementioned nonlinear systems, it can only guarantee convergence to the target state over an infinite time horizon. This limitation renders it unsuitable for applications with strict temporal constraints on convergence time, as it fails to ensure finite-time convergence. Below, we present the current research status of three types of time-constrained control methodologies.

Compared to asymptotical stabilizing control with an infinite convergence time, **finite-time control** not only ensures that state trajectories converge to equilibrium within a finite time interval but also endows the system with enhanced response speed and control precision, thus garnering extensive attention and research efforts. In the late 1990s, with the emergence of finite-time homogeneity theory (Bhat & Bernstein 1997, 2005) and the finite-time Lyapunov stability theorem (Bhat & Bernstein 1995), the finite-time control framework based on homogeneity methods experienced rapid development. Bhat et al. established a finite-time stability criterion in Bhat and Bernstein (1997) based on homogeneity theory, stating that a system is finite-time stable if it is asymptotically stable and possesses a negative homogeneity degree. However, the stability criterion in Bhat and Bernstein (1997) is only applicable to homogeneous systems. To address finite-time control for non-homogeneous systems, Hong Yiguang et al. proposed the extended homogeneity theorem (Hong et al. 2001; Hong 2002), which decomposed a non-homogeneous system into a non-homogeneous term and a homogeneous subsystem. If the original non-homogeneous system is asymptotically stable, the homogeneous subsystem is finite-time stable, and the non-homogeneous term satisfies specific conditions, the original system can be proven finite-time stable. Although significant progress has been made in finite-time control of nonlinear systems based on homogeneity or extended homogeneity theorems, deriving explicit expressions for convergence time remains challenging (Yin et al. 2017; Guo et al. 2024; Yang & Cao 2010; Mondal et al. 2018; Orlov 2004). However, homogeneous technique-based finite-time control typically requires precise system models, rendering it less applicable to practical nonlinear systems with disturbances or model uncertainties. As a finite-time control methodology free from chattering and singularity issues, the power-integration technique (Huang et al. 2005) has been widely applied to high-order nonlinear systems and nonlinear systems with uncertainties and disturbances. The power-integration finite-time control demonstrates superior performance in avoiding chattering and singularities compared to terminal sliding mode finite-time control approaches (Zheng & Li 2019; Jiang et al. 2017; Fu et al. 2015; Fan et al. 2021). Fu et al. (2017) investigated the global adaptive finite-time stabilization for a class of uncertain nonlinear systems with positive odd rational powers, considering both parameter uncertainties and unknown control coefficients. The study combined power-integration methods with logic-based switching control to address these challenges. Hong et al. (2006) examined the global finite-time stabilization of p-normal form nonlinear systems with parametric uncertainties. By integrating backstepping techniques with power-integration methods, they derived an adaptive finite-time control law in the form of continuous time-invariant feedback. In

Zhao et al. (2023a), the authors successfully designed a non-Lipschitz continuous state feedback controller for stochastic nonlinear systems using sign functions and power-integration techniques, achieving rapid finite-time control of stochastic systems. Wen et al. (2023) addressed the finite-time consensus tracking problem for high-order multi-agent systems with external disturbances and bounded control inputs. The authors developed a continuous finite-time consensus tracking protocol by combining power-integration techniques with a saturation-constrained super-twisting algorithm, ensuring tracking error convergence to zero within a finite time. For practical applications where the motion direction of system dynamics may be unknown, Zhou et al. (2023) and Wang et al. (2018) proposed robust finite-time controllers by incorporating Nussbaum functions with power-integrators.

It is noteworthy that research has demonstrated the convergence time in finite-time control depends on the system's initial states. Excessively large or unknown initial states may lead to prolonged or unpredictable convergence durations. Thus, **fixed-time control** was proposed, which can achieve faster convergence rate and better disturbance rejection properties. Moreover, the convergence time is only related to the system parameters, and is independent of the initial conditions of the system. Andrieu et al. (2008) established the homogeneous approximation theory and the bi-limit homogeneity theorem, which laid the foundation for rapid development of homogeneity-based fixed-time control techniques. The bi-limit homogeneity theorem demonstrates that a system achieves fixed-time stability when its zero-limit system possesses a negative homogeneous degree while the infinite-limit system exhibits a positive homogeneous degree. Building upon the bi-limit homogeneity theory, Tian et al. (2017) designed a second-order fixed-time state observer and investigated fixed-time output feedback control for disturbed second-order integrator systems. Tian et al. (2018) developed both fixed-time disturbance observers and tracking controllers for high-order integrator systems with unmatched disturbances, though without providing convergence time estimates. In Guo and Hu (2023a), the authors employed adaptive neural networks to construct fixed-time state observers, utilizing the bi-limit homogeneity theorem to establish a doubly-proportional indirect system for determining stability time. This approach yields an explicit expression for the convergence time upper bound with reduced conservatism. Recent applications include quadrotor tracking and formation control: Shao et al. (2021) and Mechali et al. (2022) respectively implemented fixed-time tracking and formation for disturbed quadrotors through bi-limit homogeneity theorem and output feedback methods. In recent years, power-integration techniques have been widely adopted by researchers in fixed-time control of various nonlinear systems to circumvent singularity issues commonly which encountered during fixed-time back-stepping control design. Sui et al. (2023) and Xu et al. (2023) integrated adaptive techniques with power-integration methods to investigate non-singular fixed-time tracking control for multi-input multi-output nonlinear systems with non-strict feedback structures. For stochastic nonlinear systems, Min et al. (2022) proposed an improved fixed-time Lyapunov theorem featuring a more rigorous stability proof procedure. Notably, this theorem provides a less conservative estimate of the convergence time upper bound. Furthermore, power-integration fixed-time control has been extensively applied to cooperative tracking and formation control of multi-agent systems. For instance, Min et al. (2022) studied fixed-time consensus tracking for second-order nonlinear multi-agent systems with leader-follower architecture. A fixed-time tracking controller based on power-integrator techniques was developed for each agent, enabling followers to track the leader's states within a fixed time. Ni and Shi (2021) combined nonlinear mapping techniques with power-integrator methods to design distributed fixed-time observers and radial basis function neural network-based fixed-time tracking controllers for multi-agent systems subject to output constraints, unknown control directions, unknown system dynamics, external disturbances, and input deadzones.

Research has demonstrated that the upper bound of convergence time in finite-time control depends on the system's initial conditions, whereas it is determined by the system parameters in fixed-time control systems. However, in both cases, the upper bound of convergence time cannot be arbitrarily predetermined because it is influenced by either the initial conditions or control parameters. In many practical engineering applications, there is often a critical need to arbitrarily adjust or even prescribe the convergence time of system states to meet rapid response requirements. Therefore, in recent years, **prescribed-time control** has attracted the keen interest of many researchers. This concept was first introduced by Song et al. (2017), infusing vigor into finite/fixed-time control and also giving rise to a multitude of subsequent studies. Based on different control methodologies, the existing prescribed-time control can be categorized into two approaches: scaling design and non-scaling design. The scaling design method can further be divided into state scaling and time transformation methods (Song et al. 2023). The state scaling design method utilizes the characteristic of time-varying gain functions that grow to infinity as time approaches the preset time to scale the system states. This transforms the original system's prescribed-time stabilization problem into a bounded stability issue for the scaled system states. Based on the state scaling method, Song et al. investigated the prescribed-time control problem for uncertain integral chain systems with nonlinear terms matching the control inputs and provided criteria for prescribed-time stability (Song et al. 2017). They further elaborated on the advantages of prescribed-time control over finite/fixed-time control (Song et al. 2019). Gao et al. studied the prescribed-time state feedback control problem for switched nonlinear systems (Gao et al. 2019). Wang et al. designed a time-varying feedback controller using a transformation function to achieve precise tracking control of nonlinear systems affected by parametric uncertainties (Wang & Song 2019). Holloway et al. addressed the prescribed-time estimation and output feedback control problems for linear systems in both observable and controllable canonical forms (Holloway & Krstic 2019a,b). Steeves et al solved the global prescribed-time output feedback control problem for reaction-diffusion equations (Krishnamurthy et al. 2020a) and linear Schrödinger equations (Steeves et al. 2020). Espitia et al. designed a controller with time-varying gains to ensure the prescribed-time stability of single-input linear time-invariant systems with constant input delays (Espitia & Perruquetti 2022), and further constructed an observer for linear time-invariant systems with measurement delays, ensuring observer errors converge to zero within a specified terminal time (Espitia et al. 2022). Ye et al., addressing norm systems with non-vanishing uncertainties/disturbances, introduced a smooth bi-tanh function to construct a nonlinear time-varying state feedback control scheme, and further extended it to solve the output feedback-based prescribed-time stabilization problem by employing a prescribed-time observer (Ye & Song 2023). The time transformation design method utilizes a nonlinear time transformation to map the time scale $t \in [0, T)$ to $\tau \in [0, \infty)$, achieving control design for asymptotic stability based on τ to realize prescribed-time stability based on t. Based on the time transformation design method, Yucelen et al. achieved prescribed-time control for a first-order multi-agent system (Yucelen et al. 2019). Krishnamurthy et al. proposed a prescribed-time controller using dynamic gain scaling techniques, ensuring the global prescribed-time stability of uncertain nonlinear systems. In these systems, the entire dynamical model includes state-dependent uncertainty terms that satisfy known nonlinear growth conditions (Krishnamurthy et al. 2020a). Tran et al. introduced the definition of the generalized finite-time gain function, detailed the design process of the prescribed-time control algorithm under the time transformation method, and applied it to the distributed control of networked multi-agent systems (Tran & Yucelen 2020). For strict-feedback systems with dynamics that include state-dependent nonlinear terms and uncertain parameters, Krishnamurthy et al. designed dynamic observer and controller based on the dual dynamic high-gain approach (Krishnamurthy et al. 2020b). These designs enable estimation and regulation of system states within a prescribed time. Furthermore, leveraging non-standard dynamic gain techniques and high-gain scaling parameters, they achieved prescribed-time output feedback control for more general uncertain nonlinear systems, allowing for the presence of cross-products of unknown parameters and unmeasured state variables in the entire dynamical model Krishnamurthy et al. (2021)). Professor Orlov first integrated finite-time control with prescribed-time control, proposing a finite-time varying-gain approach to achieve prescribed-time stabilization control for a class of nonlinear systems (Orlov 2022). Based on two key transformations: state scaling and time transformation, Liu et al. addressed the prescribed-time control problem for uncertain nonlinear systems with unknown control directions and gains, but the controller only operates within a preset time interval rather than the entire time domain (Liu & Liu 2023b). The non-scaling design approach employs time-varying gain functions directly in the controller design, rather than for state scaling or time transformation. Based on the non-scaling design philosophy, Zhou et al. utilized a linear time-varying control strategy to solve the parameter Lyapunov equation, achieving prescribed-time state feedback and output feedback control for both linear and nonlinear systems (Zhou 2020b,a; Zhou & Shi 2021). Li et al. (2023a) also addressed the design of unknown input observers under periodic delay output, ensuring observer errors converge to zero within the prescribed time. Pal et al. proposed a non-scaling control strategy ensuring convergence at any time and sufficient conditions for Lyapunov stability (Pal et al. 2020). However, their main results were proven to be non-rigorous (Orlov & Krstic 2022). He et al. effectively relaxed the constraints on continuous flows by capturing and utilizing the positive effects of impulses, achieving prescribed-time stabilization for nonlinear impulsive systems (He et al. (2023)). Li et al. studied the prescribed-time state feedback (Li & Krstic 2021), output feedback (Li & Krstic 2022a), and mean square bounded control (Li & Krstic 2023) problems for stochastic nonlinear systems, ensuring prescribed-time convergence in probability for stochastic nonlinear systems. According to different design ways of control schemes, prescribed-time control can be categorized into two approaches: scaling-based design and non-scaling design. At present, most existing works focus on prescribed-time control of nonlinear systems under scaling design methods, that is, employing prescribed-time tuning functions to scale system states or perform time scaling transformations. This transforms the original system's prescribed-time control problem into a bounded stability issue for the scaled system states or an asymptotic stability issue for system states under transformed time scales. In such cases, the prescribed-time functions within the controller are evidently of high order. As time approaches the preset convergence time, the computational burden of the prescribed-time functions and their derivatives increases sharply, particularly when dealing with such issue of high-order systems.

In addition to the aforementioned prescribed-time control, there is also a category of **practical prescribed-time control**. The control method based on the prescribed performance function, which is currently referred to as Prescribed Performance Control (PPC) in research, is another means of achieving output constraints. Since its initial proposal by Bechlioulis in Bechlioulis and Rovithakis (2008a,b), this method has garnered widespread attention. Its core concept can be summarized as two transformations and one equivalence. First, a prescribed performance function is designed to transform the original constrained tracking error into an unconstrained variable, namely, the transformed error. Then, a transformed error dynamics model is constructed by taking the transformed error variable as the new system state. Finally, a controller is designed to ensure that the transformed error is bounded, which is equivalent to guaranteeing that the original tracking error of the system converges within the predefined bounds, while the convergence rate is higher than the decay rate of the prescribed performance function, and the maximum overshoot is less than the initial value of the prescribed performance function. The prescribed performance function based on the exponential function is the most fundamental and widely used function. The reason why the exponential function is widely used as a prescribed performance function is that it is a sufficiently smooth function, and its derivatives of any order can be easily calculated, which facilitates the subsequent controller design. Therefore, the majority of literature related to PPC employs the exponential function to achieve output constraints (Li et al. 2020b; Wang et al. 2022e). However, this method has the issue of initial error constraints, that is, the initial value of the function must be greater than the initial value of the system tracking error. In practical systems, the initial state of the system is often not accurately obtainable, and this prerequisite cannot be guaranteed, which severely limits the applicability of the PPC method based on the exponential function. To overcome this problem, researchers have proposed some prescribed performance functions that are not dependent on initial conditions. These include the prescribed performance functions based on the hyperbolic tangent function (Bu et al. 2015) and the reciprocal of the exponential function (Xiao 2018). On the other hand, the prescribed performance functions mentioned above are all monotonically decreasing functions, which means that the prescribed performance functions will converge to the set steady-state error range in an infinite time. To accelerate the convergence rate, a novel finite-time prescribed performance function has been proposed by researchers (Sui et al. 2021b; Sun et al. 2020c). This function can rapidly converge to its terminal value within a predefined finite time and remain at that value. Since the system's tracking error is constrained by the prescribed performance function, it can also converge to the desired range within a finite time.

1.4.3 Brief Overview of State Constraints Control

State constraints are ubiquitous in practical industrial systems, and control problems within this domain have attracted growing attention from researchers in recent years. Prior to elaborating on the current research landscape of state-constrained systems, it is pertinent to review two representative functions of seminal significance. Ngo et al. (2005) introduced the Log-Barrier Lyapunov Function (LBLF) formulation, a pivotal contribution to the analysis of state-constrained systems. Tee et al. (2009) proposed the Log-Asymmetric Barrier Lyapunov Function (Log-ABLF), a generalized framework for handling asymmetric state constraints in nonlinear control systems. This formulation enhances the flexibility of barrier-based control strategies by accommodating asymmetric constraint boundaries. The proposition of Logarithmic Barrier Lyapunov Function (Log-BLF) and Asymmetric Logarithmic Barrier Lyapunov Function (Log-ABLF) has shed new light on state-constrained control of nonlinear systems. Over the past decade, a plethora of control algorithms based on BLF and ABLF have emerged, particularly flourishing in the realm of robust adaptive control. This section will systematically provide a detailed exposition on current research advancements in both output constraints and full-state constraints.

In addition to the aforementioned two classical BLFs, Xu and Jin (2013) proposed a tangent function-based BLF (Tan-BLF) for **output-constrained control** systems. Jin (2017) addressed output constraints in stochastic nonlinear systems with actuator failures by developing a fault-tolerant robust adaptive control scheme using Tangent-type Barrier Lyapunov Functions (Tan-BLF). This framework explicitly incorporates failure magnitude bounds through projection operators, while maintaining prescribed output confinement via barrier function. However, the aforementioned works are confined to handling constant output constraints. Tee et al. (2011) made a pioneering breakthrough by integrating the Logarithmic Asymmetric Barrier Lyapunov Function (Log-ABLF) with backstepping techniques, thereby extending the constant constraint paradigm to time-varying constraints for nonlinear systems. Subsequently, Qiu et al. (2015) proposed an adaptive control algorithm for lower-triangular nonlinear systems, utilizing Log-ABLF and dynamic surface control techniques. This algorithm not only circumvents the "explosion of complexity" issue inherent to traditional backstepping control strategies but also effectively addresses the time-varying constraint problems of system outputs. Meng et al. (2015) employed output-constraint transformation techniques to convert a MIMO nonlinear system with output constraints into a new "unconstrained" system. Subsequently, by integrating neural networks, they proposed an intelligent control strategy that ensured the stability of the "unconstrained" system and the boundedness of all closed-loop signals. Subsequently, Hong (2002) built upon the foundation laid in literature (Meng et al. 2016), investigating the issue of time-varying output constraints for pure-feedback nonlinear systems. To relax the restrictions on initial conditions imposed by the BLF control strategy, researchers have introduced the integral form of the barrier Lyapunov function (IBLF). Based on the Integral Barrier Lyapunov Function (IBLF), the system's state constraint issues can be directly addressed, and the problem of the initial value selection range being too small is solved.

For the problem of system output constraints, the issue of **full-state constraints control** is more universal and challenging. This is because, in addition to imposing restrictions on the system's outputs, the full-state constraint problem also places constraints on the remaining states of the system. As a widely used method for state constraints, the BLF approach achieves state-constrained control by constructing a specialized Lyapunov function tied to the states and their constraint boundaries. The term Barrier Lyapunov Function originates from its distinctive property: the BLF value approaches positive or negative infinity as the state approaches the upper or lower constraint boundaries (Ren et al. 2010; Niu & Zhao 2013; An et al. 2017). Consequently, if the initial state lies within the constraints, ensuring the boundedness of the BLF through controller design guarantees perpetual satisfaction of the state constraints. Over the past decade, significant progress has been made in BLF-based state-constrained control for nonlinear systems. For uncertain strict-feedback nonlinear systems with full-state constraints, Liu and Tong (2017), Zhang et al. (2018c), and Wu and Xie (2021) developed adaptive controllers by Log-BLF combined with Nussbaum gain techniques. A Tan-BLF was introduced in Jin (2018) to address output-constrained nonlinear systems with actuator faults. Subsequent studies Sun et al. (2020b), Gao et al. (2021), and Liu et al. (2022c) leveraged Tan-BLF to design adaptive controllers for high-order full-state constraints nonlinear systems, achieving convergence of tracking errors to arbitrarily small neighborhoods of the origin while preventing state violations. However, the methods in Li (2020), Liu and Tong (2017), Zhang et al. (2018c), Wu and Xie (2021), Jin (2018), Sun et al. (2020b), Gao et al. (2021), and Liu et al. (2022c) are only applicable to systems with symmetric state constraints. To address asymmetric state constraints, He et al. (2017) proposed an Asymmetric Barrier Lyapunov Function (ABLF), which exhibits broader applicability compared to traditional BLFs. This innovation has spurred numerous studies on ABLF-based state-constrained control (Liu et al. 2022f; Tee et al. 2009; Xue & Ou 2023). Notably, conventional BLFs (e.g., Log-BLF and Tan-BLF) indirectly achieve constraint problems by transforming state constraints into tracking error constraints, often imposing overly conservative restrictions on initial conditions (Kim & Yoo 2015; Tang et al. 2016;

Liu et al. 2017; Tee & Ge 2012). To overcome this limitation, Tee and Ge (2012) introduced an IBLF that directly enforces state constraints, thereby relaxing initial condition requirements. Building on this, Tian and Song (2023) developed an iterative adaptive control algorithm for high-order uncertain nonlinear systems using IBLF, eliminating the need for additional error dynamics constraints while ensuring signal tracking. Singh and Jain (2024) designed a collision-avoidance algorithm for multi-agent systems with time-varying constraints by integrating ABLF with backstepping control. However, it should be noted that BLF (or IBLF)-based backstepping control schemes require that virtual controls α_i must satisfy so-called feasibility conditions $-\pi_{i1} < \alpha_i < \pi_{i2}$, where π_{i1} and π_{i2} are constraint boundaries. In general, the offline optimization method is used to search for proper design parameters satisfying the feasibility conditions, which is complicated and may cause the control scheme to be inapplicable. To circumvent the "feasibility conditions" inherent in BLF-based methods, the Nonlinear Transformation Function (NTF) method was introduced (Zhang et al. 2017a). Consequently, the NTF method has garnered significant attention and widespread application in recent years (Xin et al. 2023; Wang et al. 2023a). A logarithmic composite function form of NTF was proposed in Zhang et al. (2017a). In this work, the NTF technique was employed to deal with the constrained states, while the radial basis function (RBF) neural network technique was utilized to estimate the unknown nonlinear terms of the system. Ultimately, an effective adaptive tracking control algorithm was successfully designed. For systems with both static and dynamic state constraints, Zhang et al. (2021b) proposed a unified NTF compatible with hybrid constraints. By integrating an event-triggered mechanism and dynamic surface filters, the study achieved tracking control and state constraint enforcement for multi-input-multi-output (MIMO) nonlinear systems. A ratio-type NTF was introduced in Zhao and Song (2020) to address asymmetric motion constraints in robotic systems. Building on Zhao and Song (2020) and Zhao et al. (2019a) combined the ratio-type NTF with time-varying gain techniques to guarantee finite-time transient performance and state constraints for strict-feedback nonlinear systems. To unify symmetric and asymmetric constraint management, Zhao et al. (2023b) developed a generalized nonlinear transformation. This approach ensures global boundedness of closed-loop systems while satisfying diverse state constraints.

1.5 Organization of the Book

This book proposes an adaptive constrained control framework for HOFA nonlinear systems. It addresses input constraints (managing actuator saturation), time constraints (optimizing delay and transient responses), and state constraints (ensuring safety boundaries). By integrating Lyapunov stability theory with real-time parameter estimation, the framework resolves unmodeled dynamics and time-delay effects, incorporating event-triggered control (reducing computational load) and multi-variable decoupling strategies. Validation through trajectory tracking of unmanned

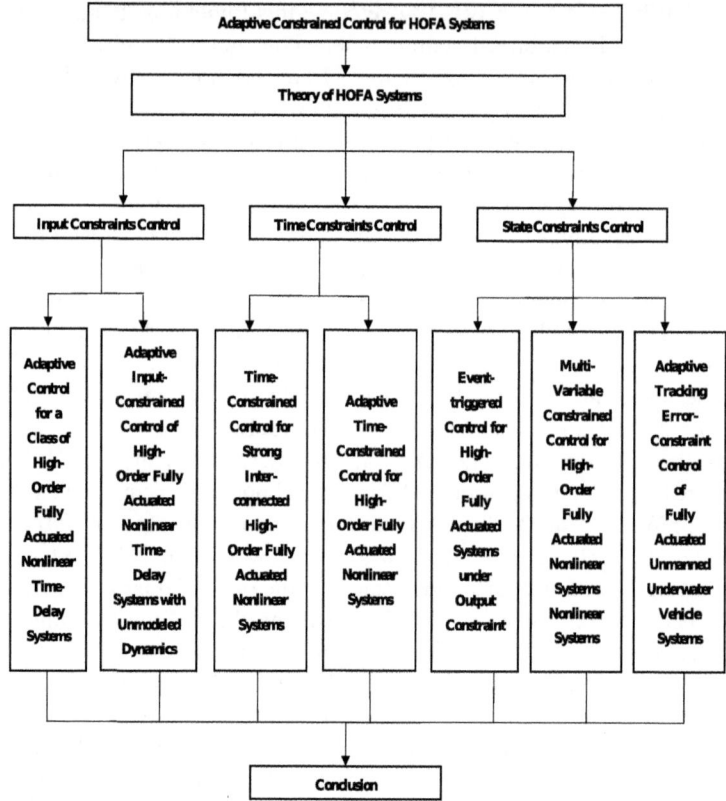

Fig. 1.1 Organization of the Book

underwater vehicles (UUVs) demonstrates robustness under hydrodynamic uncertainties. Figure 1.1 delineates a comprehensive schematic representation of the architectural framework underpinning the research content.

Chapter 2
Preliminaries

This chapter introduces some basics of the theory of FASs. The goal is to establish a foundational understanding necessary for the development and analysis of advanced control strategies in later chapters. This study establishes mathematical representations of fully actuated nonlinear system dynamics, followed by systematic analysis of adaptive control architectures for uncertainty/disturbance attenuation. Subsequent exploration addresses constrained control synthesis under practical limitations (input saturation, state constraints, temporal restrictions), concluding with formalization of fundamental mathematical constructs—including operator notation, variational inequalities, and functional analytic apparatus—essential for subsequent theoretical developments.

2.1 Fully Actuated Nonlinear Models

FASs refer to a system in which the number of independent control inputs equals the number of degrees of freedom of the system. Such systems are widely encountered in robotics, aerospace, and mechanical applications. A general nonlinear model of a FAS can be expressed as

$$Ex^{(m)} = f(x^{(0 \sim m-1)}) + B(x^{(0 \sim m-1)})u, \tag{2.1}$$

where $m \geq 1$ is an integer, $x \in \mathbb{R}^n$ is the system state vector, $(x^{(0 \sim m-1)})^T = [x, \dot{x}, \cdots, x^{(m-1)}]$, $u \in \mathbb{R}^r$ is the control input vector, $f(x^{(0 \sim m-1)}) \in \mathbb{R}^n$ is a smooth vector function, and $B(x^{(0 \sim m-1)}) \in \mathbb{R}^{n \times r}$ is a matrix-valued function.

© The Author(s) 2026
C. Hua et al., *Adaptive Constrained Control for High-Order Fully Actuated Nonlinear Systems*, https://doi.org/10.1007/978-981-95-0962-1_2

Definition 2.1 (Duan (2021a)) For any $x^{(i)} \in \Omega_i \subset \mathbb{R}^n$, $i = 0, 1, \ldots, m-1$, if the following condition holds:

$$\text{rank } B\left(x^{(0 \sim m-1)}\right) = r = n, \quad \forall t \geq 0, \tag{2.2}$$

then the system (2.1) is referred to as a FAS, and $\Omega = \Omega_0 \times \Omega_1 \times \cdots \times \Omega_{m-1}$ is called the attraction domain of system (2.1).

Let $A_i \in \mathbb{R}^{r \times r}$, for $i = 0, 1, \ldots, m-1$, be a set of given design matrices. Then, the following controller is designed as

$$u = -B^{-1}(x^{(0 \sim m-1)})\left(A_{0 \sim m-1} x^{(0 \sim m-1)} + f(x^{(0 \sim m-1)}) - v\right), \tag{2.3}$$

which is applied to the FAS (2.1), yields the following constant linear closed-loop system:

$$Ex^{(m)} + A_{0 \sim m-1} x^{(0 \sim m-1)} = v. \tag{2.4}$$

Upon expansion, this linear system can also be written as:

$$Ex^{(m)} + A_{m-1} x^{(m-1)} + \cdots + A_1 \dot{x} + A_0 x = v. \tag{2.5}$$

Due to the freedom in selecting the matrices A_i, $i = 0, 1, \ldots, m-1$, closed-loop systems with desired dynamic characteristics can be achieved by appropriately choosing these matrices. A parametric method for determining such matrices has been proposed in Duan (2021a).

2.1.1 Linear Systems

Duan (2024) consider a linear system in the following state-space form:

$$\dot{x} = Ax + Bu, \tag{2.6}$$

where $x \in \mathbb{R}^n$ and $u \in \mathbb{R}^r$ are the state vector and input vector, respectively, and $A \in \mathbb{R}^{n \times n}$ and $B \in \mathbb{R}^{n \times r}$ are the coefficient matrices.

The linear system (2.6) is controllable if and only if it can be converted equivalently into the following multi-order global FAS:

$$
\begin{bmatrix} x_1^{(\mu_1)} \\ x_2^{(\mu_2)} \\ \vdots \\ x_\eta^{(\mu_\eta)} \end{bmatrix} = \begin{bmatrix} L_1 \left(x_k^{(0 \sim \mu_k - 1)} \big|_{k=1 \sim \eta} \right) \\ L_2 \left(x_k^{(0 \sim \mu_k - 1)} \big|_{k=1 \sim \eta} \right) \\ \vdots \\ L_\eta \left(x_k^{(0 \sim \mu_k - 1)} \big|_{k=1 \sim \eta} \right) \end{bmatrix} + \tilde{B} u,
\tag{2.7}
$$

where $\eta \geq 1$ is an integer, μ_k, $k = 1, 2, \ldots, \eta$, are a set of integers, $x_k \in \mathbb{R}^{r_k}$, $k = 1, 2, \ldots, \eta$, are a set of state vectors, with r_k, $k = 1, 2, \ldots, \eta$, being a set of distinct integers satisfying (11). Further, $L_k(\cdot)$, $k = 1, 2, \ldots, \eta$, are a set of linear functions, and $\tilde{B} \in \mathbb{R}^{r \times r}$ is a constant square upper-triangular matrix with diagonal elements all being.

2.1.2 Strict-Feedback Systems

In the historical development of control theory, the investigation of strict-feedback nonlinear systems has consistently remained a central research focus, signifying a crucial paradigm shift from linear system theory to nonlinear system theory. These systems are widely found in various engineering applications, including robotics, aircraft systems, and chemical processes. Characterized by their distinctive structural configuration, such systems exhibit a unique property where the derivative of each state variable functionally depends on preceding state variables, thereby establishing a hierarchical control architecture. This particular structure not only presents theoretical challenges in system analysis, but also creates extensive design possibilities for control strategy formulation. The structural representation of strict-feedback nonlinear systems is illustrated as follows:

$$
\begin{cases} \dot{x}_i(t) = x_{i+1}(t) + f_i\left(x_1(t), x_2(t), \ldots, x_i(t)\right), & i = 1, \cdots, n-1 \\ \dot{x}_n(t) = u(t) + f_n\left(x_1(t), x_2(t), \ldots, x_n(t)\right), \\ y(t) = x_1(t). \end{cases}
\tag{2.8}
$$

For generalized high-order strict-feedback systems, Academician Duan Guang-Ren has developed an iterative equivalent transformation methodology to systematically convert them into HOFA system models (Duan 2021b). The structured conversion process is algorithmically outlined as follows:

$$\begin{cases} x_1 = z, \\ x_2 = \dot{z} - f_1(z), \\ x_{k+1} = z^{(k)} - \sum_{i=1}^{k} f_i^{(k-i)} \left(z^{(0\sim i-1)} \right), \\ k = 2, 3, \ldots, n-1 \end{cases} \tag{2.9}$$

where $z^{(0\sim i-1)} = \left[z, \dot{z}, \ldots, z^{(i-1)} \right]$.

Based on the aforementioned transformation methodology, the strict-feedback nonlinear system model (2.8) can be converted into the following HOFA system model.

$$z^{(n)} = u + \sum_{i=1}^{n} f_i^{(k-i)} \left(z^{(0\sim i-1)} \right). \tag{2.10}$$

If the controller is designed as

$$\begin{cases} u = - \left(A_{0\sim n-1} z^{(0\sim n\ 1)} + u^* \right), \\ u^* = \sum_{i=1}^{k} f_i^{(k-i)} \left(z^{(0\sim i-1)} \right) - v. \end{cases} \tag{2.11}$$

The HOFA system can be transformed into a linear time-invariant closed-loop system with the desired eigenstructure.

$$z^{(n)} + A_{0\ n-1} z^{(0\sim n-1)} = v. \tag{2.12}$$

In the formulation, v is the intermediate control law to be designed for addressing certain nonlinear problems, while $A_{0\sim n-1}$ denotes an arbitrary matrix that ensures the stability of $\Phi(A_{0\sim n-1})$.

2.2 Adaptive Control Framework

In practical applications, the system dynamics often suffer from parametric uncertainties or unmodeled dynamics. Adaptive control addresses such challenges by dynamically adjusting control parameters based on real-time system behavior. A typical adaptive control framework consists of two main components: a control law ensuring stability under nominal conditions and an adaptation law for estimating uncertain parameters.

To demonstrate the basic design method of adaptive control, consider the following FAS with uncertain parameters:

$$x^{(m)} = G^T(x^{(0\sim m-1)})\theta + f(x^{(0\sim m-1)}) + B(x^{(0\sim m-1)})u, \tag{2.13}$$

where $\theta \in \mathbb{R}^r$ and $G(x^{(0\sim m-1)}) \in \mathbb{R}^{r\times n}$ are uncertain parameter vector and sufficiently smooth vector function, respectively, other variables and functions are defined in (2.1).

In order to establish the control law of the FAS (2.13) with uncertain parameters, we need the following preparations.

When the matrix $A^{0\sim m-1}$ is chosen that the matrix $\Phi(A^{0\sim m-1})$ is stable, it follows from the well known Lyapunov Theorem that there exists a positive definite matrix $P(A^{0\sim m-1})$ satisfying

$$\Phi^T(A^{0\sim m-1})P(A^{0\sim m-1}) + P(A^{0\sim m-1})\Phi(A^{0\sim m-1}) \leq -I_{mn},$$

$$P(A^{0\sim m-1}) = [P_1 \quad P_2 \quad \cdots \quad P_m], \; P_i \in \mathbb{R}^{mn\times n}. \tag{2.14}$$

Theorem 2.1 *Let $A_i \in \mathbb{R}^{mn\times n}$ be a set of matrices such that the matrix $\Phi(A^{0\sim m-1})$ is stable, and $P_L(A^{0\sim m-1}) = P(A^{0\sim m-1})[0, I_n]^T = P_m$. Then, for any unknown constant vector $\theta^T \in \mathbb{R}^r$, there exists an adaptive control law*

$$\dot{\hat{\theta}} = G(x^{(0\sim m-1)})P_L^T(A^{0\sim m-1})x^{(0\sim m-1)},$$

$$u = -B^{-1}(x^{(0\sim m-1)})\left(A^{0\sim m-1}x^{(0\sim m-1)} + G^T(x^{(0\sim m-1)})\hat{\theta} + f(x^{(0\sim m-1)})\right), \tag{2.15}$$

for the system (1) such that the states of the closed-loop system satisfy $\lim_{t\to+\infty} x^{(0\sim m-1)}(t) = 0$ and $\|\theta - \hat{\theta}\|$ is bounded.

Proof First, it follows from (2.13) and (2.15) that

$$x^{(m)} + A^{0\sim m-1}x^{(0\sim m-1)} = G^T(x^{(0\sim m-1)})\tilde{\theta},$$

$$\dot{x}^{(0\sim m-1)} = \Phi(A^{0\sim m-1})x^{(0\sim m-1)} + [0_{(m-1)n}, G^T x\tilde{\theta}]^T, \tag{2.16}$$

where $\tilde{\theta} = \theta - \hat{\theta}$ is the estimation error of θ with the estimated value $\hat{\theta}$. Then, the Lyapunov function candidate is defined as follows:

$$V = (x^{(0\sim m-1)})^T P(A^{0\sim m-1})x^{(0\sim m-1)} + \tilde{\theta}^T\tilde{\theta}, \tag{2.17}$$

where $P(A^{0 \sim m-1})$ is the solution to (2.14). The time derivative of V along the trajectories of the closed-loop system is given by

$$\dot{V} = (\dot{x}^{(0 \sim m-1)})^T P(A^{0 \sim m-1}) x^{(0 \sim m-1)}$$

$$+ (x^{(0 \sim m-1)})^T P(A^{0 \sim m-1}) \dot{x}^{(0 \sim m-1)} - 2\dot{\tilde{\theta}}^T \tilde{\theta}$$

$$= (x^{(0 \sim m-1)})^T (\Phi^T P + P\Phi) x^{(0 \sim m-1)}$$

$$+ 2\left[(x^{(0 \sim m-1)})^T P_L(A^{0 \sim m-1}) G^T x - \dot{\tilde{\theta}}^T\right] \tilde{\theta}. \tag{2.18}$$

In addition, it follows from (2.14), (2.15) and (2.18) that

$$\dot{V} \leq -\|x^{(0 \sim m-1)}\|^2. \tag{2.19}$$

It can be thus obtained from Barbalat's Lemma with $W = V$ that $\|\tilde{\theta}\|$ is bounded, and $\lim_{t \to +\infty} x^{(0 \sim m-1)}(t) = 0$. This proof is finished.

2.3   Constrained Control Methodologies

The constraints in nonlinear systems primarily encompass three categories: **input constraints**, **output and full-state constraints**, and **time constraints**.

2.3.1   Input Constraints

Control strategies for input constraints are generally classified into six types: **input saturation, input dead-zone, input delay, hysteresis, input quantization** , and **actuator fault constraints**.

   Input saturation is a nonlinear phenomenon in control systems, where the control signals generated by the controller (e.g., voltage, current, torque) are forcibly truncated due to physical actuator limitations (e.g., maximum voltage, maximum thrust), resulting in actual inputs failing to reach their theoretically computed values. Its mathematical representation is given by:

$$u(t) = Sat(v(t)) = \begin{cases} v_{max}, & v(t) \geq v_{max} \\ v(t), & v_{min} < v(t) < v_{max} \\ v_{min}, & v(t) \leq v_{min} \end{cases}, \tag{2.20}$$

where $Sat(\cdot) \in R$ is an asymmetric saturation nonlinear characteristic, $v(t) \in R$ is input signal generated by the controller and $u(t)$ is the control input. The saturation nonlinearity $u(t)$ is addressed via an approximate nonlinear function decomposition as follows by Wen et al. (2011):

$$u(t) = Sat(v(t)) = k_1(v) + k_2(v), \qquad (2.21)$$

where $k_1(v) = v_{max} tanh(\frac{v(t)}{v_{max}})$, $k_2(v) = Sat(v(t)) - k_1(v)$. By the Mean Value Theorem, there exists a constant $\mu (0 < \mu < 1)$, such that:

$$k_1(v) = k_1(v_0) + k_{\mu 1}(v - v_0), \qquad (2.22)$$

where $k_{\mu 1} = \frac{\partial k_1(v)}{\partial v} |_{v=v_\mu}$, $v_\mu = \mu + (1 - \mu)v_0$. By setting $v_0 = 0$, one can obtain:

$$k_1(v) = k_{\mu 1} v, \qquad (2.23)$$

then in the controller design, parameter $k_{\mu 1}$ is adapted to use the adaptive law. $k_2(v)$ is a bounded function in time and its bound can be obtained as

$$| k_2(v) = Sat(v(t)) - k_1(v) | \le v_{max}(1 - tanh(1)) = D_1. \qquad (2.24)$$

Li et al. (2017b) eliminated input saturation by incorporating an auxiliary variable

$$\dot{\tilde{h}} = -\tilde{h} + u(t) - v(t), \qquad (2.25)$$

in the final step of the backstepping framework for high-order nonlinear systems.

$$z_n = x_n - \alpha_{n-1} - \tilde{h}, \qquad (2.26)$$

where z_n is the system state, α_{n-1} is an virtual controller.

Building on this work, Gao et al. (2017) extended the application of auxiliary variables

$$\begin{cases} \dot{\eta}_1 = -\delta_1 \eta_1 + \eta_2, \\ \dot{\eta}_i = -\delta_i \eta_i + \eta_{i+1}, & i = 2, \ldots n - 1 \\ \dot{\eta}_n = -\delta_n \eta_n + u(t) - v(t), \end{cases} \qquad (2.27)$$

to every step of the controller design,

$$\begin{cases} z_1 = x_1 - y_r - \eta_1, \\ z_i = x_i - \alpha_{i-1} - \eta_i, & i = 2, \ldots n, \end{cases} \qquad (2.28)$$

where $\delta_1 \ge \frac{1}{2}$ and $\delta_i \ge 1 (i = 2, \ldots n)$ are designed parameters. $z_i (i = 1, \ldots n)$ are the system states, $\alpha_{i-1} (i = 1, \ldots n - 1)$ are virtual controllers. It effectively compensates for the effects of saturation and further analyzes the error stability issues caused by input saturation.

Input dead zone, a quintessential nonlinear phenomenon in control systems, describes the scenario where actuators (e.g., motors, hydraulic valves) fail to produce effective outputs when the amplitude of input signals (e.g., voltage, current, torque) lies below a specific threshold, termed the deadband width. This behavior fundamentally originates from static friction (stiction), mechanical backlash, or resolution limitations in sensors/actuators. Its mathematical representation is given by:

$$u(t) = D(v(t)) = \begin{cases} m_r(v(t) - b_r), & v(t) \geq b_r \\ 0, & b_l < v(t) < b_r \\ m_l(v(t) - b_l), & v(t) \leq b_l \end{cases} \tag{2.29}$$

where $D(\cdot) \in \mathbb{R}$ is the input of dead-zone, m_r and m_l represent respectively the left slope and right slope of the dead zone characteristic. At the same time, the parameters b_l and b_r represent respectively the left and right breakpoints of the input nonlinearity, and $b_r - b_l$ stands for the dead-zone width. $v(t)$ denotes the controller-output command signal, while $u(t)$ represents the effective signal actually applied to the actuator. To mitigate the adverse effects of dead-zone inputs in practical engineering applications, $u(t)$ is addressed via an approximate nonlinear function decomposition as follows:

$$u(t) = D(v(t)) = m(t)v(t) + q(t), \tag{2.30}$$

where

$$m(t) = \begin{cases} m_r, & v(t) \geq b_r \\ 0, & b_l < v(t) < b_r \\ m_l, & v(t) \leq b_l \end{cases} \tag{2.31}$$

$$q(t) = \begin{cases} -m_r b_r, & v(t) \geq b_r \\ -mv(t), & b_l < v(t) < b_r \\ -m_l b_l, & v(t) \leq b_l \end{cases} \tag{2.32}$$

Clearly, there exist positive constants $\underline{m}, \bar{m}, \bar{q}$, such that $\underline{m} \leq m(t) \leq \bar{m}$, where $\underline{m} = \min(m_r, m_t), \bar{q} = \max(m_r b_r, m_l b_l)$.

Input delay refers to the latency that occurs between a system receiving an input signal and its actual commencement of processing or responding to that signal. This delay may be caused by processing time in hardware, software, or communication links, and is commonly observed in fields such as control systems, electronic devices, and real-time interactive systems. Its mathematical representation is given by:

$$u(t) = v(t - \tau), \tag{2.33}$$

where τ represents the input time delay. In order to address the input delay existing in the system, Wu et al. (2025) introduced the following Pade approximation method:

$$\mathcal{L}[v(t - \tau)] = e^{-\tau s}\mathcal{L}[v(t)]\frac{e^{\frac{-\tau s}{2}}}{e^{\frac{\tau s}{2}}}\mathcal{L}[v(t)] \approx \frac{1 - \frac{\tau s}{2}}{1 + \frac{\tau s}{2}}\mathcal{L}[v(t)], \qquad (2.34)$$

where $\mathcal{L}[v(t)]$ represents the Laplace transform of $v(t)$, and s represents the Laplace variable. Define a new variable $x_{n+1} = [\frac{1 - \frac{\tau s}{2}}{1 + \frac{\tau s}{2}}]v + v$ satisfying

$$2\mathcal{L}[v(t)] = \mathcal{L}[x_{n+1}] + \frac{\tau s}{2}\mathcal{L}[x_{n+1}]. \qquad (2.35)$$

On the basis of the inverse Laplace transform, (2.35) can be rewritten as

$$\dot{x}_{n+1} = \frac{4}{\tau}u - \frac{2}{\tau}x_{n+1}. \qquad (2.36)$$

In contrast to Wu et al. (2025), the delay $\tau(t)$ in Xia et al. (2022) is upper time-varying bounded with $\tau(t) < \tau_{max}$. Xia et al. (2022) establishes the stability of time-delay systems through the construction of a Lyapunov-Krasovskii function.

$$V(x) = \frac{1}{2}\sum_{i=1}^{n}z_i^2 + \frac{p}{2(1 - \beta)}\int_{t-\tau(t)}^{t}|u(\phi)|^2\,d\phi$$

$$+ \frac{p}{(1 - \beta)}\int_{t-\tau(t)}^{t}(\int_{s}^{t}|u(\phi)|^2\,d\phi)ds, \qquad (2.37)$$

where z_i is an error variable, p is a positive constant and β satisfies $|\dot{\tau}(t)| < \beta < 1$.

Hysteresis is a prevalent nonlinear phenomenon in control systems, mechanical engineering, and materials science, arising from the dependence of a system's output not only on its instantaneous input but also on the historical trajectory of the input. In nonlinear control, the P-I model (Hua & Li 2015) is formulated as follows:

$$\begin{aligned}
&u(t) = p_0 v(t) - d[v](t),\\
&d[v](t) = \int_0^D p(r)F_r[v](t)dr,\\
&F_r[v](0) = f_r(v(0), 0),\\
&F_r[v](t) = f_r(v(t), F_r[v](t_m)), \quad t_m < t \le t_{m+1}, 0 \le m \le M - 1\\
&f_r(k, w) = max(k - r, min(k + r, w)),
\end{aligned} \qquad (2.38)$$

where $p_0 = \int_0^D p(r)dr$, $p(r)$ is a given density function satisfying $p(r) \ge 0$ with $\int_0^{+\infty} rp(r)dr < +\infty$; D is a constant so that the density function $p(r)$ vanishes at the threshold parameter r. $v(t)$ is the input of the hysteresis model; $0 = t_0 < t_1 < \ldots < t_M = t_E$ is the partition of $[0, t_E]$ and M is a positive integer; Hua and Li (2015) effectively compensated for unknown P-I hysteresis models via an adaptive variable structure control approach, without requiring the construction of a hysteresis inverse.

Another common hysteresis model is the Bouc-Wen (B-W) model. This model is commonly employed in structural dynamics to characterize nonlinear restoring forces, such as force-displacement relationships in dampers or materials subjected to cyclic loading conditions. By incorporating a hysteretic variable, the model captures memory effects of the system, where the current state depends not only on instantaneous inputs but also on the historical loading path. B-W model is formulated as follows:

$$
\begin{aligned}
&u(t) = \textstyle\prod(v(t)) = \rho\xi v(t) + (1-\rho)\xi h(t) = k_1 v(t) + k_2 h(t),\\
&\dot{h}(t) = \dot{v}(t) - \alpha \mid \dot{v}(t) \parallel h(t) \mid^{\sigma-1} h(t) - \theta \dot{v}(t) \mid h(t) \mid^{\sigma}, \qquad h(t_0) = 0,\\
&\dot{h}(t) = \dot{v}(t) - \alpha \mid \dot{v}(t) \mid h(t) \mid - \theta \dot{v}(t) \mid h(t) \mid = \dot{v}(t) - f(h, \dot{v}),\\
&f(h, \dot{v}) = 1 - sign(\dot{v}(t))\alpha h(t) - \theta \mid h(t) \mid .
\end{aligned}
$$
$$(2.39)$$

In control systems, input quantization refers to the process of converting continuous analog input signals (e.g., sensor measurements) into discrete numerical values with finite precision.

The uniform quantizer is modeled as Zhou et al. (2019)

$$
Q_u = \begin{cases} u_i, & u_i - \frac{l}{2} \le u < u_i + \frac{l}{2},\\ 0, & -u_0 \le u < u_0,\\ -u_i, & -u_i - \frac{l}{2} \le u < -u_i + \frac{l}{2}, \end{cases}
\qquad (2.40)
$$

where $u_0 = \frac{l}{2}$ determines the length of the dead zone, $u_1 = l$, $u_{i+1} = u_i + l, i \in N_{1,\infty}$, l is the length of the quantization interval. The uniform quantizer satisfies $\mid Q_u - u \mid \le \frac{l}{2}$.

The hysteresis-uniform quantizer is modeled as Li et al. (2024a)

$$
Q_{h-u} = \begin{cases} & u_i - \frac{l}{2} - h < \mid u \mid \le u_i - \frac{l}{2} + h\\ & \quad and \quad \dot{u} < 0, \quad or\\ u_i sign(u), & u_i + \frac{l}{2} - h < \mid u \mid < u_i + \frac{l}{2} + h\\ & \quad and \quad \dot{u} > 0, \quad or\\ & u_i - \frac{l}{2} + h \le \mid u \mid \le u_i + \frac{l}{2} - h\\ 0, & -u_0 - h \le u \le u_0 + h,\\ Q_{hu}(u(t^-)), & \dot{u} = 0, \end{cases}
\qquad (2.41)
$$

where $u_0 = \frac{l}{2}$ determines the length of the dead zone, $u_1 = l$, $u_{i+1} = u_i + l, i \in N_{1,\infty}$, l is the length of the quantization interval. $h = 0.5kl$ is the hysteresis width parameter, and $0 < k \le 1$ is the hysteresis percentage. It is easy to verify that the error of the hysteresis quantizer is bounded, where $\mid Q_u - u \mid \le \frac{l}{2} + h$.

The hysteresis quantizer is described as Wang and Wang (2022)

$$
Q_{hu} = \begin{cases}
u_i \, \text{sign}(u), & \begin{aligned} &\frac{u_i}{1+\delta} < |u| \le u_i, \; \dot{u} < 0, \\ &\text{or} \quad u_i < |u| \le \frac{u_i}{1-\delta}, \; \dot{u} > 0, \end{aligned} \\[2ex]
\bar{u}_i \, \text{sign}(u), & \begin{aligned} &u_i < |u| \le \frac{u_i}{1-\delta}, \; \dot{u} < 0, \\ &\text{or} \quad \frac{u_i}{1-\delta} < |u| \le \frac{u_i(1+\delta)}{1-\delta}, \; \dot{u} > 0, \end{aligned} \\[2ex]
0, & \begin{aligned} &0 \le |u| \le \frac{u_i}{1+\delta}, \; \dot{u} < 0, \\ &\text{or} \quad \frac{u_i}{1+\delta} < |u| \le u_i, \; \dot{u} > 0, \end{aligned} \\[2ex]
Q_h(u(t^-)), & \text{otherwise,}
\end{cases}
\tag{2.42}
$$

where $\bar{u}_i = u_i(1+\delta)$, $u_i = \rho^{1-i} u_{min}$, $i \in N_{1,\infty}$. $u_{min} > 0$ represents the dead-zone range of $Q(hu)$ with parameter $\delta = \frac{1-\rho}{1+\rho}$, $0 < \rho < 1$. Respectively. Q_{hu} can be written as

$$
Q_{hu} = H(u)u + L(u),
\tag{2.43}
$$

when $-u_{min} < u < u_{min}$, $H(u) = 1$, $L(u) = -u$, otherwise, $H(u) = \frac{H(u)}{u}$ and $L(u) = 0$. The control coefficient $H(u)$ and input quantization error $L(u)$ satisfy $H_{min} < H(u) < H_{max}$, and $|L(u)| \le \bar{L}$, where $H_{min} = 1 - \delta$ $H_{max} = 1 + \delta$, $\bar{L} = u_{min}$.

The logarithmic quantizer is modeled as Xing et al. (2016)

$$
Q_{lu} = \begin{cases}
u_i sign(u), & \frac{u_i}{1+\rho} < |u| \le \frac{u_i}{1-\rho}, \\
0, & |u| \le \frac{u_{min}}{1+\rho},
\end{cases}
\tag{2.44}
$$

where $u_i = \iota^{1-i}$ with $i = 1, 2, ..$ and $\iota = \frac{1-\rho}{1+\rho}$ with $0 < \rho < 1$. The quantization error of quantizer Q_{lu} is bounded by $|Q_{lu} - u| \le \delta$, where $\delta \ge \frac{l}{2}$ is the maximum quantization interval length.

Actuator fault constraints refer to a set of restrictive conditions and strategies designed in control systems to address potential actuator failures (e.g., lock-in-place, loss of effectiveness, and drift). These constraints ensure system stability and safety under fault conditions. The mathematical characterization is formulated as follows:

$$
u(t) = k_1(t)v(t) + k_2(t),
\tag{2.45}
$$

where $k_1(t) \in (0, 1)$ denotes the unknown time-varying actuator efficiency factor, $k_2(t)$ is a bounded time-varying bias fault satisfying $|k_2(t)| < \bar{k}_2$, and $\bar{k}_2$ is a positive constant. $v(t) \in \mathbb{R}$ is the control signal to be designed and $u(t)$ is the quantized control input of the system.

2.3.2 Time Constraints

In control systems, time constraints refer to specific limitations or conditions that a system must satisfy along the temporal dimension. These constraints directly influence the system's dynamic performance, stability, real-time responsiveness, and the task execution feasibility. Time constraints are primarily categorized into **finite-time control**, **fixed-time control**, and **prescribed-time control**.

Finite-time control is fundamentally aimed at driving system states to converge to a desired equilibrium point or a tracking target within a finite time interval (as opposed to asymptotic convergence), while guaranteeing tunability or predictability of the convergence duration. It is widely applied in high-dynamic scenarios such as robotics, aerospace, and power systems.

Classical methods for achieving finite-time control include sliding surface design, power-integration control, and homogeneity theory. In practical applications, appropriate approaches should be selected based on system characteristics (such as degree of nonlinearity, disturbance types, and system order), while balancing convergence time, robustness, and implementation complexity.

Terminal sliding mode control is a state-of-the-art control method that combines the robustness of sliding mode control with finite-time convergence properties. Its core lies in the design of nonlinear sliding mode surfaces so that the system state converges to an equilibrium point in finite time. Conventional sliding mode control uses a linear sliding mode surface $s = \dot{e} + \lambda e$ to achieve asymptotic convergence of states. To achieve finite-time exact tracking, Mao et al. (2025) defined the integral sliding mode variables by

$$s = e + \lambda \int_0^t sign(e(\tau))\mathrm{d}\tau, \tag{2.46}$$

where e is the tracking error ($e = x_1 - y_d$, y_d is the reference signal), and $sign(\cdot)$ is the standard sign function. The Lyapunov function $V = \frac{1}{2}s^2$ is then constructed and its derivative is calculated:

$$\dot{V} = s\dot{s} \leq -aV^{\frac{1}{2}}, \tag{2.47}$$

where $a > 0$, it indicates that the state of the system reaches the sliding mode surface $s = 0$ in a finite time and further converges to the equilibrium point along the sliding mode surface in finite time. The convergence time satisfies

$$T \leq \frac{2}{a}V^{\frac{1}{2}}(0). \tag{2.48}$$

Power-integral control accelerates the finite-time convergence of system states by introducing power terms and integral operations, while suppressing disturbances and mitigating the accumulation of historical errors. Incorporating power terms of state variables into the control law enhances system convergence speed through nonlinear feedback. The introduction of power terms accelerates convergence as the system approaches the equilibrium point. An integral term serves to accumulate historical error information, dynamically adjusting control inputs to effectively suppress steady-state errors caused by disturbances or model uncertainties. By combining Lyapunov functions with homogeneity theory, rigorous finite-time convergence guarantees are established, ensuring system states converge to the equilibrium point or track reference trajectories within a predefined time.

Liu and Liu (2023a) constructed the Lyapunov function

$$V = \frac{1}{2}x_1^2 + \sum_{i=2}^{n} \int_{-\beta_{i-1}z_{i-1}^{q_{i-1}}}^{x_i} (s^{\frac{1}{q_{i-1}}} + \beta_i^{\frac{1}{q_{i-1}}} z_{i-1})^{2+(i-2)\sigma-q_{i-1}} ds, \quad (2.49)$$

where β_i are virtual laws, $z_i = x_i^{\frac{1}{q_{i-1}}} + \beta_{i-1}^{\frac{1}{q_{i-1}}} z_{i-1}$ and its derivatives are calculated:

$$\dot{V} \le -cV^{\frac{r}{2}}, \quad (2.50)$$

then x_i converges to zero before finite time $\frac{2}{c(1-\frac{r}{2})} V^{1-\frac{r}{2}}(x(0))$. And it is rigorously proven via finite-time Lyapunov theory.

As proposed in Bhat and Bernstein (1997): For system $\dot{x} = f(x,t)$, if there exists a $\mathbb{C}^1$ function $V(x) \ge 0$ such that

$$\dot{V}(x) \le -kV^q(x), \quad (2.51)$$

where $k > 0, 0 < q < 1$, then the closed-loop system is finite-time (FT) stable and the settling time is calculated by

$$T := \frac{1}{k(1-q)} V^{1-q}(x(0)). \quad (2.52)$$

As proposed in Shen and Xia (2008): For system $\dot{x} = f(x,t)$, if there exists a $\mathbb{C}^1$ function $V(x) \ge 0$ such that

$$\dot{V}(x) \le -k_1 V^q(x) + k_2 V(x), \quad (2.53)$$

where $k_1 > 0$, $k_2 > 0$ and $0 < q < 1$, then the closed-loop system is practical semi-global FT stable and the settling time is calculated by

$$T := \frac{1}{k_2(1-q)} \ln \left(1 - \frac{k_1}{k_2} V^{1-q}(x(0))\right). \quad (2.54)$$

By designing a discontinuous feedback controller and selecting a nonsmooth Lyapunov function such that its time derivative satisfies (2.51) or (2.53), the finite-time asymptotic convergence of system states can be rigorously proven, as per the extended Lyapunov stability criteria.

As proposed in Wang et al. (2017): For system $\dot{x} = f(x, t)$, if there exists a $\mathbb{C}^1$ function $V(x) \geq 0$ such that

$$\dot{V}(x) \leq -kV^q(x) + \eta, \tag{2.55}$$

where $k > 0$, $0 < \eta < \infty$ and $0 < q < 1$, then the closed-loop system is practical semi-global FT stable and the settling time is calculated by

$$T := \frac{1}{k\theta(1-q)}(V^{1-q}(x(0)) - (\frac{\eta}{k(1-\theta)})^{\frac{1-q}{q}})), \tag{2.56}$$

where $0 < \theta < 1$ is a constant.

As proposed in Yu et al. (2018): For system $\dot{x} = f(x, t)$, if there exists a $\mathbb{C}^1$ function $V(x) \geq 0$ such that

$$\dot{V}(x) \leq -k_1 V^q(x) + k_2 V(x) + \eta, \tag{2.57}$$

where $k_1 > 0$, $k_2 > 0$, $0 < \eta < \infty$ and $0 < q < 1$, then the closed-loop system is practical FT stable and the settling time is calculated by

$$T := max \left\{ \frac{\ln \frac{k_2\theta V^{1-q}(x(0))+k_1}{k_1}}{k_2\theta(1-q)}, \frac{\ln \frac{k_2 V^{1-q}(x(0))+\theta k_1}{\theta k_1}}{k_2(1-q)} \right\}, \tag{2.58}$$

where $0 < \theta < 1$ is a constant.

Fixed-time control (FXT) is a methodology that guarantees system states converge to an equilibrium point or a tracking target within a fixed upper-bounded time interval independent of initial conditions.

Classical fixed-time control methods achieve explicit constraints on convergence time by incorporating nonlinear feedback, time-varying gains, or fixed-time sliding mode, combined with fixed-time Lyapunov functions. The design necessitates a trade-off among robustness, control input magnitude, and implementation complexity.

The core of fixed-time sliding mode control lies in designing a sliding surface incorporating dual-power terms, which ensures rapid convergence from a distance while avoiding singularity issues.

$$s = \dot{e} + \beta_1 e^{\frac{q_1}{p_1}} + \beta_2 e^{\frac{q_2}{p_2}}, \tag{2.59}$$

where e is the tracking error ($e = x_1 - y_d$, y_d is the reference signal), β_1, $\beta_2 > 0$, $q_1 < p_1$, $q_2 > p_2$, and q_i, p_i are positive odd numbers. When the error is substantial, the dominant term $\beta_1 e^{\frac{q_1}{p_1}}$ accelerates the convergence rate from regions far from the equilibrium point. When the error is small, the dominant term $\beta_2 e^{\frac{q_2}{p_2}}$ prevents the singularity issues inherent in terminal sliding mode designs thus ensuring fixed-time convergence. Finally, by constructing a Lyapunov function and invoking the fixed-time Lyapunov stability theorem, it is rigorously proven that the system states converge within a fixed time.

As proposed in Polyakov (2012): For system $\dot{x} = f(x, t)$, if there exists a $\mathbb{C}^1$ function $V(x) \geq 0$ such that

$$\dot{V}(x) \leq -(\alpha V^p(x) + \beta V^q(x))^k, \tag{2.60}$$

where $\alpha > 0, \beta > 0, p > 0, q > 0, k > 0$, and $pk < 1, qk > 1$, then the closed-loop system is FXT stable and the settling time is bounded by

$$T := \frac{1}{\alpha^k(1 - pk)} + \frac{1}{\beta^k(qk - 1)}. \tag{2.61}$$

As proposed in Polyakov (2020): For system $\dot{x} = f(x, t)$, if there exists a $\mathbb{C}^1$ function $V(x) \geq 0$ such that

$$\dot{V}(x) \leq -\alpha V^p(x) - \beta V^q(x), \tag{2.62}$$

where $\alpha > 0, \beta > 0, p = 1 - \frac{1}{2\gamma}, q = 1 + \frac{1}{2\gamma}, \gamma > 0$, then the closed-loop system is FXT stable and the settling time is bounded by

$$T := \frac{\pi \gamma}{\sqrt{\alpha \beta}}. \tag{2.63}$$

As proposed in Zuo and Tie (2014): For system $\dot{x} = f(x, t)$, if there exists a $\mathbb{C}^1$ function $V(x) \geq 0$ such that

$$\dot{V}(x) \leq -\alpha V^{2 - \frac{p}{q}}(x) - \beta V^{\frac{p}{q}}(x), \tag{2.64}$$

where $\alpha > 0, \beta > 0, q > p > 0$, both q and p are odd integers, then the closed-loop system is FXT stable and the settling time is bounded by

$$T := \frac{q\pi}{2\sqrt{\alpha\beta}(q - p)}. \tag{2.65}$$

As proposed in Zuo and Tie (2016): For system $\dot{x} = f(x, t)$, if there exists a $\mathbb{C}^1$ function $V(x) \geq 0$ such that

$$\dot{V}(x) \leq -k_1 V^{\frac{m}{n}}(x) - k_2 V^{\frac{p}{q}}(x), \tag{2.66}$$

where $k_1 > 0, k_2 > 0, q > p > 0, m > n > 0$, and q and p, m and n are odd integers, then the close-loop system is FXT stable and the settling time is bounded by

$$T := \frac{1}{k_1}\frac{n}{m-n} + \frac{1}{k_2}\frac{q}{q-p}. \tag{2.67}$$

Prescribed-time control is a strategy that enables users to directly specify the convergence time, where system states are guaranteed to converge to the equilibrium within a user-specified time T.

The core idea of prescribed-time control is to compress the system dynamics into a prescribed-time through time-varying gain or timescale transformation. For example, a time scaling function $\mu(t) = \frac{T}{T-t}$ is introduced to force the state to converge by designing a special Lyapunov function and its derivatives so that the system gain tends to infinity at $t \to T$.

As proposed in Song et al. (2017): For system $\dot{x} = f(x,t)$, consider a time-varying function $\mu(t) = \frac{T}{T-t}$, if there exists a $\mathbb{C}^1$ function $V(x) \geq 0$ satisfies

$$\dot{V}(x) \leq -2k\mu(t)V + \frac{\mu(t)}{4\theta}d(t)^2, \tag{2.68}$$

for unknown perturbation $d(t)$ and positive numbers k, θ, then $V(t)$ is bounded for $t \in [0, T)$.

As proposed in Zou et al. (2023): Consider a smooth function

$$\zeta_0(x) = \begin{cases} 0, & x \in (-\infty, 0], \\ e^{-1/x}, & x \in (0, +\infty), \end{cases} \quad x \in \mathbb{R} \tag{2.69}$$

and define

$$\zeta_a(t, T) = \frac{\zeta_0(T-t)}{\zeta_0(T-t) + \zeta_0(t)}, \tag{2.70}$$

where $T > 0$ is a parameter determined by the controller designer. A time-dependent scaling function $\zeta(t, T, \varepsilon) \in \mathbb{R}$ is constructed as

$$\zeta(t, T, \varepsilon) = \frac{1}{\zeta_a(t, T) + \varepsilon} = \frac{\zeta_0(T-t) + \zeta_0(t)}{(1+\varepsilon)\zeta_0(T-t) + \varepsilon\zeta_0(t)}. \tag{2.71}$$

where $0 < \varepsilon \ll 1$.

If there exits a Lyapunov function $V(t)$ which satisfies

$$\dot{V}(t) = -\left(k + \frac{\dot{\zeta}(t, T, \varepsilon)}{\zeta(t, T, \varepsilon)}\right)V, \quad V_0 = V(0), \tag{2.72}$$

where $k > 0$ is a positive constant, and $\xi(t, t_f, \varepsilon)$ is defined as (2.71), then $V(t)$ converges to the region $V \le V_0$ with a predefined time and then approaches to zero asymptotically.

Considering a time-varying function defined over the entire time interval as follows (Luo et al. 2024):

$$\mu(t) = \begin{cases} \left(\dfrac{T^*}{T^* + t_0 - t} \right)^q, & t \in [t_0, T^* + t_0 - \varepsilon), \\ \left(\dfrac{T^*}{\varepsilon} \right)^q, & t \in [T^* + t_0 - \varepsilon, +\infty), \end{cases} \tag{2.73}$$

where $q \ge \max\{n, 2\}$ is any user-defined real number, n denotes the system order, T^* is any physically allowable finite time pre-specified by the designer, and ε is a small constant satisfying $0 < \varepsilon < T^*$. It is noted that the defined function μ is continuous, monotonically increasing, and bounded for $t \in [t_0, +\infty)$ with $\mu(t_0) = 1$, $\mu(t) \to \left(\frac{T^*}{\varepsilon} \right)^q$ as $t \to (T^* + t_0 - \varepsilon)^-$, and $\mu(t) = \left(\frac{T^*}{\varepsilon} \right)^q$ for all $t \ge T^* + t_0 - \varepsilon$.

Let $\psi(t)$ be an unknown bounded function and $\|\psi\|_{[t_0,T]} = \sup_{\tau \in [t_0,T]} |\psi(\tau)|$, and let $V(t) : [t_0, +\infty) \to \mathbb{R}^+ \cup \{0\}$ be a continuously differentiable function, then

1. If

$$\dot{V}(t) \le -k\mu(t)V - \frac{2q}{T^*}\mu(t)^{\frac{1}{q}}V + \frac{\psi(t)}{\mu(t)}, \tag{2.74}$$

for $t \in [t_0, T^* + t_0 - \varepsilon)$, it holds that

$$V(t) \le \frac{\zeta_1(t)}{\mu(t)^2}V(t_0) + \frac{\|\psi\|_{[t_0,T]}}{k\mu(t)^2}, \tag{2.75}$$

where k is a finite positive constant, $\zeta_1(t) = \exp\left(-\frac{kT^*}{q-1} \left(\left(\frac{T^*}{T^* + t_0 - t} \right)^{q-1} - 1 \right) \right)$ is a monotonically decreasing function with $\zeta_1(t_0) = 1$ and $\zeta_1(t) \to 0$ as $t \to (T^* + t_0 - \varepsilon)^-$ and $\varepsilon \to 0$;

2. If

$$\dot{V}(t) \le -\bar{k}\mu V(t) - \frac{2q}{T^*}\mu^{\frac{1}{q}}V(t) + \frac{\psi(t)}{\mu}, \tag{2.76}$$

for $t \in [T^* + t_0 - \varepsilon, +\infty)$, it holds that

$$V(t) \le \frac{\zeta_2(t)\zeta_1(T^* + t_0 - \varepsilon)}{\mu^2}V(t_0)$$
$$+ \frac{\zeta_2(t)\|\psi\|_{[t_0, T^* + t_0 - \varepsilon]}}{k\mu^2} + \frac{\|\psi(t)\|_{[T^* + t_0 - \varepsilon, t]}}{\bar{k}\mu^2}, \tag{2.77}$$

where $\bar{k}$ is a finite positive constant, and $\zeta_2(t) = \exp\left(-\bar{k}\left(\frac{T^*}{\varepsilon}\right)^q (t - T^* - t_0 + \varepsilon)\right)$ is a monotonically decreasing function. It holds that $\zeta_2(T^* + t_0 - \varepsilon) = 1$ and $\zeta_2(t) \to 0$ as $t \to +\infty$ and $\varepsilon \to 0$.

As proposed in Song et al. (2023): For system $\dot{x} = f(x, t)$, consider a time-varying function $\mu(t) = \frac{T}{T-t}$, if there exists a $\mathbb{C}^1$ function $V(x) \geq 0$, $V : [0, T) \to [0, \infty)$ satisfies

$$\dot{V}(x) \leq -k\mu(t)V + |d(t)|, \tag{2.78}$$

where $d(t)$ is an unknown and bounded perturbation and k is a positive number k, then $V(t)$ is bounded for $t \in [0, T)$ and $\lim_{t \to T} V(t) = 0$.

As proposed in Hua et al. (2021): For system $\dot{x} = f(x, t)$, consider a time-varying function $\mu(t) = \frac{T}{T-t}$, if there exists a $\mathbb{C}^1$ function $V(x) \geq 0$, $V : [0, T) \to [0, \infty)$ satisfies

$$V(t) = V_1(x(t)) + V_2(\tilde{\Theta}(t)), \quad \forall x \in \mathbb{R}^n, \tag{2.79}$$

$$\alpha_1(\|x\|) \leq V_1(x) \leq \alpha_2(\|x\|), \quad \forall x \in \mathbb{R}^n, \tag{2.80}$$

$$\alpha_3(\|\tilde{\Theta}\|) \leq V_2(\tilde{\Theta}) \leq \alpha_4(\|\tilde{\Theta}\|), \tag{2.81}$$

$$\dot{V}(t) \leq -c\mu(t)V_1(x(t)), \quad \forall t \in [0, T_p), \tag{2.82}$$

then the equilibrium of the system $\dot{x} = f(x, t)$ is globally prescribed-time stable.

2.3.3 State Constraints

In nonlinear control, the significance of output constraints and state constraints is particularly pronounced due to the complex dynamic characteristics of nonlinear systems, which may exhibit behaviors such as multiple equilibrium points, limit cycles, and chaotic motion. The proper design of these constraints is critical to ensuring system safety, stability, and practical feasibility.

The outputs of nonlinear systems typically represent physical quantities that are either directly measurable or require strict tracking (e.g., temperature, position, velocity). By enforcing output constraints, the system can be guaranteed to operate within desired bounds, thereby avoiding performance degradation caused by overshoot or steady-state errors. However, state variables in nonlinear systems (e.g., velocity, acceleration, internal energy) may not be directly measurable, and their violation of constraints could lead to catastrophic system failure. Full-state constraints address this challenge by restricting all state variables, ensuring the global safety of internal dynamics and mitigating risks arising from unmeasurable variables. This approach transforms the problem of constraint enforcement into a boundedness analysis of the nonlinear transformation functions governing system behavior.

In the study of output constraints and state constraints, the constrained problem is typically transformed into a boundedness analysis of **barrier Lyapunov functions (BLFs)** and **nonlinear transformation functions**. Specifically, the design of BLFs ensures that system trajectories remain within predefined constraints by encoding the barrier properties into the Lyapunov function structure. BLFs are generally categorized into five forms, denoted as: **Log-BLF** (Ngo et al. 2005), **Log-ABLF** (Tee et al. 2009), **Tan1-BLF** and **Tan2-BLF** (Zhao et al. 2018), and **IBLF** (Yuan et al. 2023). The **nonlinear transformation functions** further map the constrained state space to an unconstrained domain, enabling the application of standard control design tools while preserving safety guarantees.

Log-BLF As proposed in Ngo et al. (2005), a Log-BLF with the following structure is introduced:

$$V = \frac{1}{2} \log \frac{K_b^2}{K_b^2 - s^2}, \tag{2.83}$$

where s denotes the constrained variable and K_b represents the constraint boundaries. If the initial value of the constrained variable $s(0)$ is defined within a subinterval of the open interval

$$D := \{s \in R \mid -K_b < s < K_b\}, \tag{2.84}$$

then as s approaches the boundary of D, Log-BLF tends to infinity, such that:

$$s \to K_b \quad or \quad s \to -K_b \Rightarrow V \to \infty. \tag{2.85}$$

If the initial value satisfies $s(0) \in D' \subsetneq D$, and Log-BLF remains bounded on $[0, \infty)$, then $s(t) \in D$ for all $t \geq 0$. This guarantees that the constrained variable remains strictly within the prescribed constraints. The methodology is inherently restricted to symmetric-form state constraint problems.

Log-ABLF As proposed in Tee et al. (2009), a Log-ABLF with the following structure is introduced:

$$V = \frac{q(s)}{p} \log \frac{K_b^p}{K_b^p - s^p} + \frac{1 - q(s)}{p} \log \frac{K_a^p}{K_a^p - s^p}, \quad q(\cdot) = \begin{cases} 1, & \cdot > 0, \\ 0, & others, \end{cases} \tag{2.86}$$

where s denotes the constrained variable, K_b and $-K_a$ represent the constraint boundaries. $p \in \mathbb{Z}^+$ is an even integer such that $p > n$, n denotes the system order. If the initial value of the constrained variable $s(0)$ is defined within a subinterval of the open interval

$$D_1 := \{s \in R \mid -K_a < s < K_b\}, \tag{2.87}$$

then as s approaches the boundary of D_1, Log-ABLF tends to infinity, such that:

$$s \to K_b \quad or \quad s \to -K_a \Rightarrow V \to \infty, \tag{2.88}$$

if Log-ABLF remains bounded over the time domain $[0, \infty)$, then $s \in D_1$, thereby ensuring that the constrained variable operates strictly within the asymmetric constraints.

Tan1-BLF In addition to the two aforementioned classical BLFs, a tangent function-based BLF (denoted as Tan1-BLF) was proposed in Xu and Jin (2013) , formulated as follows:

$$V = \frac{K_b^2}{\pi} \tan(\frac{\pi s^2}{K_b^2}). \tag{2.89}$$

Tan2-BLF A tangent function-based BLF (denoted as Tan2-BLF) was proposed in Zhao et al. (2018) , formulated as follows:

$$V = s \tan(\frac{\pi}{2K_b}s), \tag{2.90}$$

where s denotes the constrained variable and K_b represents the constraint boundaries. If the initial value satisfies $s(0) \in D' \subsetneqq D$, Tan1-BLF and Tan2-BLF remain bounded on $[0, \infty)$, then $s(t) \in D$ for all $t \geq 0$.

It should be noted that the aforementioned Log-BLF, Log-ABLF, Tan1-BLF and Tan2-BLF all transform the constraint problem of the state x_i into a constraint problem of the errors $s_1 = x_1 - y_d$, $s_i = x_i - \alpha_i$, where y_d is the reference signal and α_i is the virtual signal. Thus, in order to impose direct constraints on the states x_i, Yuan et al. (2023) proposed the following IBLF:

$$V_z = \int_0^z \frac{\rho k_c^2}{k_c^2 - (\rho + y_d)^2} d\rho, \tag{2.91}$$

where k_c is the constraint function, and $z = x_1 - y_d$ denotes the tracking error. By employing such IBLF, the state is directly constrained rather than the virtual error.

Nonlinear Transformation Function

$$\xi_i = \frac{x_i}{(F_{ia} + x_i)(F_{ib} - x_i)}, \tag{2.92}$$

where x_i denotes the system state, F_{ib} and $-F_{ia}$ represent the constraint boundaries. When the initial condition $x_i(0)$ satisfies $x_i(0) \in D_i$, where

$$D_i := \{x_i \in R \mid -F_{ia} < x_i < F_{ib}\}, \tag{2.93}$$

when x_i approaches the boundary of the open interval D_i, ζ_i tends to infinity. Thus, for any initial value $x_i(0) \in D_i$, if the controller design ensures the boundedness of ζ_i, then $x_i \in D_i$ will naturally hold. Consequently, the problem of full-state constraints in the system is transformed into ensuring the boundedness of the nonlinear transformation function ζ_i.

2.4 Mathematical Background

This section summarizes the essential mathematical tools and definitions that are used throughout the book. We first introduce the stability concepts, followed by a review of useful stability lemmas and mathematical inequalities.

2.4.1 Stability Definitions

Consider the following nonautonomous system:

$$\dot{x}(t) = f(t, x(t)), \quad x(t_0) = x_0, \quad t \geq t_0 \geq 0, \tag{2.94}$$

where $x(t) \in \mathbb{R}^n$ is the system state; t_0 is the initial time; $f(\cdot) : \mathbb{R}_{\geq t_0} \times \mathbb{R}^n \to \mathbb{R}^n$ is piecewise continuous with respect to t and locally Lipschitz continuous with respect to $x(t)$. Moreover, if $f(t, 0) = 0$ for all $t \geq t_0$, then $x(t) = 0$ is an equilibrium of system (2.94).

Definition 2.2 (Khalil and Grizzle (2002)) For the nonautonomous system (2.94), if for any constant $\varepsilon > 0$ and $t_0 \geq 0$, there always exists a constant $\delta = \delta(\varepsilon, t_0) > 0$ such that

$$\|x_0\| < \delta \quad \Rightarrow \quad \|x(t)\| < \varepsilon, \quad \forall t \geq t_0, \tag{2.95}$$

then the equilibrium point $x(t) = 0$ is said to be stable. Furthermore, if for any constant $\varepsilon > 0$, there always exists a constant $\delta = \delta(\varepsilon) > 0$, independent of the initial time t_0, such that (2.95) holds, then the equilibrium point $x(t) = 0$ is said to be uniformly stable.

Definition 2.3 (Khalil and Grizzle (2002)) If the equilibrium point $x(t) = 0$ of system (2.94) is stable and there exists a positive constant $c = c(t_0)$ such that

$$\|x(t)\| \to 0 \quad \text{as} \quad t \to \infty, \quad \forall \|x_0\| < c, \tag{2.96}$$

then the equilibrium point $x(t) = 0$ is said to be asymptotically stable. Furthermore, if the equilibrium point $x(t) = 0$ is uniformly stable and there exists a constant c, independent of the initial time t_0, such that (2.96) holds, then the equilibrium point $x(t) = 0$ is said to be uniformly asymptotically stable.

Definition 2.4 (Khalil and Grizzle (2002)) If the equilibrium point $x(t) = 0$ of system (2.94) is stable and satisfies

$$\|x(t)\| \to 0 \quad \text{as} \quad t \to \infty, \quad \forall x_0 \in \mathbb{R}^n, \tag{2.97}$$

then the equilibrium point $x(t) = 0$ is said to be globally asymptotically stable. Furthermore, if the equilibrium point $x(t) = 0$ is uniformly stable and, for any constants $\delta > 0$ and $\varepsilon > 0$, there exists a constant T, independent of the initial time t_0, such that

$$\|x(t)\| < \varepsilon, \quad \forall \|x_0\| < \delta, \quad \forall t \geq t_0 + T, \tag{2.98}$$

then the equilibrium point $x(t) = 0$ is said to be uniformly global asymptotic stable.

Definition 2.5 (Shen and Huang (2012)) If the equilibrium point $x(t) = 0$ of system (2.94) is asymptotically stable and, for any initial state $x_0 \in \Phi_\varepsilon := \{x \in \mathbb{R}^n : \|x\| \leq c(t_0)\}$ with some constant $c(t_0) > 0$, it holds that

$$\|x(t)\| = 0, \quad \forall t \geq T(x_0), \tag{2.99}$$

where $T(x_0)$ is a positive constant, then the equilibrium point $x(t) = 0$ is said to be finite-time stable. In particular, if $\Phi_\varepsilon = \mathbb{R}^n$, then the equilibrium point $x(t) = 0$ is said to be global finite-time stable.

Definition 2.6 (Polyakov (2012)) If the equilibrium point $x(t) = 0$ of system (2.94) is (global) finite-time stable and there exists a constant $T_{\max}$ such that

$$T(x_0) \leq T_{\max}, \tag{2.100}$$

then the equilibrium point $x(t) = 0$ is said to be (global) fixed-time stable.

Consider the following nonlinear system:

$$\dot{x}(t) = f\big(t, x(t), u\big), \quad t \geq t_0, \tag{2.101}$$

where $x(t) \in \mathbb{R}^n$ denotes the system state, $u \in \mathbb{R}^m$ represents the control input, and $t_0 \geq 0$ is the initial time. The function $f : \mathbb{R}_{\geq t_0} \times \mathbb{R}^n \times \mathbb{R}^m \to \mathbb{R}^n$ is piecewise continuous with respect to t and locally Lipschitz continuous with respect to x. Moreover, if $f(t, 0, u) = 0$ for all $t \geq t_0$ and $u \in \mathbb{R}^m$, then $x(t) = 0$ is called an equilibrium point of the nonautonomous system (2.101).

Definition 2.7 (Ning et al. (2023b)) For any given constant $T_p > 0$, if there exists a control input u such that the equilibrium point $x(t) = 0$ of the nonlinear system (2.101) is stable and, for any initial condition $x_0 \in \Phi_\varepsilon := \{x \in \mathbb{R}^n : \|x\| \leq c(t_0)\}$

with some constant $c(t_0) > 0$, the following conditions are satisfied:

$$\lim_{t \to (t_0+T_p)^-} x(t) = 0 \quad \text{and} \quad x(t) = 0, \quad \forall t \geq t_0 + T_p, \tag{2.102}$$

then the equilibrium point $x(t) = 0$ is said to be prescribed-time stable, and T_p is referred to as the prescribed time. Moreover, if $\Phi_\varepsilon = \mathbb{R}^n$, the equilibrium point $x(t) = 0$ is said to be global prescribed-time stable.

Definition 2.8 (Khalil and Grizzle (2002)) A continuous function $\psi(\gamma)$: $[0, \alpha) \to [0, +\infty)$ is said to belong to class K function if it is strictly increasing and satisfies $\psi(0) = 0$. Moreover, if $\alpha = +\infty$ and $\psi(\gamma) \to +\infty$ as $\gamma \to +\infty$, then ψ is said to belong to class K_∞ function.

2.4.2 Stability Lemmas and Inequalities

Several standard stability lemmas and inequalities are frequently employed in analysis, which are shown as follows:

Lemma 2.1 (Khalil and Grizzle (2002)) *For system (2.94), let $D \subset \mathbb{R}^n$ be a domain containing the origin. If there exists a continuously differentiable function $V(x(t)) : D \to \mathbb{R}$ satisfying $V(0) = 0$, $V(x(t)) > 0$ for all $x(t) \in D \setminus \{0\}$, and*

$$\dot{V}(x(t)) \leq 0, \quad \forall x(t) \in D, \tag{2.103}$$

then the equilibrium point $x(t) = 0$ is stable. Moreover, if

$$\dot{V}(x(t)) < 0, \quad \forall x(t) \in D \setminus \{0\}, \tag{2.104}$$

then the equilibrium point $x(t) = 0$ is said to be asymptotically stable.

Lemma 2.2 (Bhat and Bernstein (2000)) *For system (2.94), if there exists a continuously differentiable positive definite function $V(x(t))$ satisfying*

$$\dot{V}(x(t)) < -kV^q(x(t)), \tag{2.105}$$

where $k > 0$ and $0 < q < 1$ are constants, then the equilibrium point $x(t) = 0$ is finite-time stable. Moreover, the settling time T satisfies

$$T := \frac{1}{k(1-q)} V^{1-q}(x(0)). \tag{2.106}$$

Lemma 2.3 (Polyakov (2012)) *For system (2.94), if there exists a continuously differentiable positive definite function $V(x(t))$ satisfying*

$$\dot{V}(x(t)) < -kV^q(x(t)) - cV^p(x(t)), \tag{2.107}$$

where $k > 0$, $c > 0$, $0 < q < 1$, and $p > 1$ are constants, then the equilibrium point $x(t) = 0$ is fixed-time stable. Furthermore, the settling time T satisfies

$$T \leq \frac{1}{k(1-q)} + \frac{1}{c(p-1)}. \tag{2.108}$$

Lemma 2.4 (Orlov (2020)) *If a uniformly continuous function $\phi(t) : \mathbb{R}_{\geq 0} \to \mathbb{R}$ has a convergent and bounded improper integral*

$$\lim_{t \to +\infty} \int_0^t \phi(\tau) \, d\tau,$$

then $\phi(t) \to 0$ as $t \to +\infty$.

Lemma 2.5 (Duan (2021a)) *For any $\mu > 0$, there exists a set of matrices $A_i \in \mathbb{R}^{r \times r}$, $i = 0, 1, \ldots, n-1$ satisfying the following condition:*

$$\mathrm{Re}\lambda_i \left(\Phi \left(A^{0 \sim n-1} \right) \right) < -\frac{\mu}{2}, \quad i = 1, 2, \ldots, nr. \tag{2.109}$$

Lemma 2.6 (Duan (2021a)) *Suppose that $A \in \mathbb{R}^{n \times n}$ satisfies:*

$$\mathrm{Re}\lambda_i (A) < -\frac{\mu}{2}, i = 1, 2, \ldots, n, \tag{2.110}$$

where $\mu > 0$, then there exists a positive definite matrix $P \in \mathbb{R}^{n \times n}$ satisfying:

$$A^T P + PA \leq -\mu P. \tag{2.111}$$

Therefore, for any $\mu > 0$, there exists a positive definite matrix $P (\cdot) = P \left(A^{0 \sim n-1} \right)$ satisfying:

$$\Phi^T \left(A^{0 \sim n-1} \right) P (\cdot) + P (\cdot) \Phi \left(A^{0 \sim n-1} \right) \leq -\mu P (\cdot). \tag{2.112}$$

Lemma 2.7 (Young's Inequality) *If $x \in \mathbb{R}$ and $y \in \mathbb{R}$, then the following conclusions holds:*

$$xy \leq \frac{\Xi^v}{v} |x|^v + \frac{1}{w \Xi^w} |y|^w, \tag{2.113}$$

where $\Xi > 0$, $v > 0$, $w > 0$, and $(v - 1)(w - 1) = 1$.

Lemma 2.8 (Li and Yang (2016)) *For any positive continuous function $\varepsilon(t)$, the following inequality holds*

$$0 \le |b| - \frac{b^2}{\sqrt{b^2 + \varepsilon^2(t)}} < \varepsilon(t),$$

where b denotes any positive constant.

Part I
Adaptive Input-Constrained Control of High-Order Fully Actuated Nonlinear Time-Delay System

Chapter 3
Adaptive Control for a Class of High-Order Fully Actuated Nonlinear Time-Delay Systems

Building upon the basic models of fully actuated systems introduced in the previous chapter, this chapter extends the discussion to adaptive control strategies tailored for HOFA nonlinear systems with time-delay characteristics. Initially, the strict-feedback nonlinear time-delay system is transformed into a fully actuated system using fully actuated system theory. The uncertain time-delay terms are assumed to be bounded by the product of the system state's absolute value and a nonlinear function with unknown parameters. Leveraging HOFA system methodologies, a continuous adaptive controller is developed. The proposed controller is proven to ensure the system's asymptotic stability. Finally, two numerical examples are presented to validate the effectiveness of the theoretical findings.

3.1 Overview

HOFA system theory has garnered significant attention recently due to its effectiveness and convenience in addressing control problem. Introduced in Duan (2020d), HOFA system approaches highlight the limitations of first-order system methods based on state-space model, propose the concept of HOFA systems, and demonstrate their advantages in controller design. The work Duan (2020d) established that, with an appropriate nonlinear state feedback controller, a linear closed-loop system with a desired eigenstructure can be achieved. It also provided complete parametric representations for the closed-loop eigenvectors and the feedback law. Subsequently, Duan (2020b) explored the controllability of general dynamic systems using HOFA system methodologies. The concept of fully-measured systems for nonlinear dynamics was first introduced in Duan (2020c), where an observability canonical form for nonlinear systems was established. The author of Duan (2020c) demonstrated that any observable linear system and any nonlinear system equivalent to the observable canonical form can be transformed into a high-order fully-measured

© The Author(s) 2026
C. Hua et al., *Adaptive Constrained Control for High-Order Fully Actuated Nonlinear Systems*, https://doi.org/10.1007/978-981-95-0962-1_3

system. Building on this foundation, the definition of a completely observable general dynamic system was proposed. A general procedure for converting a nonlinear system into a fully actuated system was presented in Duan (2021a). Subsequently, Duan (2021b) proposed a recursive solution method to transform generalized strict-feedback systems into high-order fully actuated systems. In Duan (2021c), Duan introduced robust stabilization and tracking control schemes for single HOFA systems with nonlinear uncertainties, leveraging Lyapunov stability theory. Similarly, adaptive stabilization and tracking control schemes for high-order fully actuated systems with parametric uncertainties were developed in Duan (2021d), also based on Lyapunov stability theory. In Duan (2021e), a robust adaptive control strategy was developed for HOFA systems with both nonlinear uncertainties and time-varying unknown parameters. The work in Duan (2021f) addressed state feedback stabilization for HOFA systems with deterministic disturbances and output feedback control for systems with dynamic disturbances. Leveraging the unique characteristics of HOFA systems, Duan (2021g) proposed a framework for the controllability of general dynamical control systems, representing them as a combination of a controllable subsystem described by the HOFA model and an additional uncontrollable or supplementary subsystem. Building on HOFA system methodologies, Duan (2022b) explored the optimal control problem for general dynamical systems, while Duan (2022a) resolved the tracking control problem for HOFA systems with slow time-varying disturbances using a generalized PID control scheme. Control strategies for HOFA discrete-time systems were presented in Duan (2022c). Collectively, these contributions form the foundation of HOFA system theory. Recently, the HOFA system theory approach was extended to uncertain nonlinear impulsive systems (Huang et al. 2024), discrete-time strict-feedback systems (Xu 2024), pseudo pure-feedback nonlinear systems (Zhang et al. 2025). In addition, event-triggered control (Meng et al. 2022a), adaptive control (Wu et al. 2024), fault tolerance control (Cai et al. 2023a) and finite-time control approaches were applied to stabilization HOFA systems. However, the stabilization problem for time-delay systems using high-order fully actuated approaches remains unsolved and will be addressed in this chapter.

Time-delay phenomena are prevalent in real-world systems, including network control, cyber-physical systems, rolling systems, biological systems, ironmaking blast furnace systems, chemical processes, and communication networks. Based on the location of the time-delay terms, time-delay problems can be classified into three categories: input delays (Krstic 2009; Karafyllis & Krstic 2011; Zhu & Jiang 2014), state delays (Fridman & Shaked 2002; Ibrir 2011; Zhang et al. 2016, 2018b; Bekiaris-Liberis & Krstic 2010; Lin & Zhang 2019), and combined state and control input delays (Sipahi et al. 2012; Bekiaris-Liberis & Krstic 2012). The Lyapunov-Krasovskii (LK) and Lyapunov-Razumikhin (LR) methods are well-established and essential tools for controller design and stability analysis in time-delay systems (Lin & Zhang 2019). For instance, the global asymptotic control problem for chains of integrators with input delays was addressed in Mazenc et al. (2003) through the design of a saturation controller. Robust output feedback control for strict-feedback nonlinear time-delay systems was resolved in Hua et al. (2005), while the adaptive tracking problem for a class of nonlinear time-delay systems with unknown dead-zone inputs was explored in Hua et al. (2008b). The robust control of uncertain nonlinear time-delay systems using the Takagi-Sugeno approach was investigated in Hua et al. (2008a). Stability analysis and H_∞ control for a class of nonlinear time-delay systems were studied in Yang and Wang (2014). Adaptive fuzzy output feedback tracking control for uncertain nonlinear switched systems with time delays and unmodeled dynamics was considered in Hua et al. (2017a). Additionally, Liu and Zhang (2019) examined input-to-state stability for general nonlinear time-delay systems subject to delay-dependent impulse effects. The global state regulation problem for time-delay nonlinear systems with unknown control directions was addressed in Pongvuthithum et al. (2017). Using the dynamic gain technique, Zhang et al. (2016), Zhang et al. (2018b), and Lin and Zhang (2019) investigated the global stabilization of a class of nonlinear cascade systems with unknown time delays and unstable zero dynamics. However, the global adaptive control of nonlinear time-delay systems using high-order fully actuated system methodologies remains an open and challenging problem. This chapter investigates the global adaptive control of nonlinear systems with time delays using high-order fully actuated system methodologies.

3.2 Problem Formulation of Nonlinear Time-Delay Systems

In this section, we introduce preliminaries and problem formulation. Consider the following strict-feedback uncertain nonlinear system with time-delays,

$$\dot{x}_1 = x_2 + f_1\left(x_1, x_1(t-d), \theta\right),$$

$$\vdots$$

$$\dot{x}_i = x_{i+1} + f_i\left(x_1, \cdots, x_i, x_1(t-d), \cdots, x_i(t-d), \theta\right),$$

$$\vdots$$

$$\dot{x}_n = u + f_n\left(x_1, \cdots, x_n, x_1(t-d), \cdots, x_n(t-d), \theta\right), \tag{3.1}$$

where $x_i \in \mathbb{R}$, $i = 1, \ldots, n$ and $u \in \mathbb{R}$ are system state variables and control input respectively; $f_i(\cdot) \in \mathbb{R}$, $i = 1, \ldots, n$ are smooth functions with $f_i(0) = 0$; d is a positive unknown constant; $\theta \in \mathbb{R}^r$ is an uncertain parameter vector. It should be noted that only x_1 and its derivatives can be used for controller design.

Remark 3.1 Different from general nonlinear systems based on state space expressions, the system state variables $x_1, \ldots, x_n$ can be measured by sensors, the system (3.1) only state x_1 and its derivatives can be obtained by measurement or calculation and used for controller design. For example, robotic arms, quadrotors, etc. Their positions and speeds can all be obtained through sensors. Therefore, we turned to the HOFA system approaches to achieve the asymptotically stability, since it is relatively simple and straightforward to controller design.

To transform the system (3.1) into a fully actuated nonlinear system, the state transition is constructed as follows:

$$x_1 = z,$$
$$x_2 = \dot{z} - h_1(z, z(t-d), \theta),$$
$$\vdots$$
$$x_i = z^{(i-1)} - h_{i-1}(z^{(0 \sim i-2)}, z^{(0 \sim i-2)}(t-d), \ldots, z^{(0)}(t-(i-1)d), \theta),$$
$$\vdots$$
$$x_n = z^{(n-1)} - h_{n-1}(z^{(0 \sim n-2)}, z^{(0 \sim n-2)}(t-d), \ldots, z^{(0)}(t-(n-1)d), \theta),$$
$$\text{(3.2)}$$

where the nonlinear functions $h_i(\cdot)$, $i = 1, \ldots, n$ with time-delays is described as

$$h_1(\cdot) = f_1(z, z(t-d), \theta),$$
$$\vdots$$
$$h_i(\cdot) = \dot{h}_{i-1}(\cdot) + f_i(z^{(0 \sim i-1)}, z^{(0 \sim i-1)}(t-d), \ldots, z^{(0)}(t-id), \theta),$$
$$\vdots$$
$$h_n(\cdot) = \dot{h}_{n-1}(\cdot) + f_n(z^{(0 \sim n-1)}, z^{(0 \sim n-1)}(t-d), \ldots, z^{(0)}(t-nd), \theta). \text{(3.3)}$$

Let $h(\cdot) = h_n(\cdot)$, then it follows directly that

$$z^{(n)} = u + h\left(z^{(0 \sim n-1)}, z^{(0 \sim n-1)}(t-d), \ldots, z^{(0)}(t-nd), \theta\right),$$
$$z^{(0 \sim n-1)}(s) = v(s), \qquad s \in [-nd, 0], \text{(3.4)}$$

where $v(s) \in \mathbb{R}^n$ is a continuous vector function defined on $[-nd, 0]$. According to Definition 2.1, it is evident that the system (3.4) qualifies as a standard fully actuated nonlinear system.

Objective The primary objective of this chapter is to develop a continuous adaptive controller for the nonlinear time-delay system (3.1), using fully actuated system methodologies, to ensure that the system state x_1 and its derivatives asymptotically converge to zero.

Remark 3.2 The state space method's first-order model emphasizes the integrity and solution of the system state, while the HOFA system model focuses on control design, specifically the acquisition of control variables (Duan 2020b). Fully actuated systems can be derived either by modeling using fundamental physical laws like Newton's Law, Momentum (Moment) Theorem, Lagrange's Equation, Kirchhoff's Law, etc. (Duan 2020c), or by eliminating and upgrading underactuated systems (Duan 2020d).

Remark 3.3 Using the state transition described in (3.2), the general nonlinear system (3.1) is transformed into the fully actuated system (3.4). Consequently, the control objective for system (3.1) is reformulated as achieving asymptotic stability for the fully actuated system (3.4). Notably, since the nonlinear functions $f_i(\cdot)$ are smooth and satisfy $f_i(0) = 0$, the asymptotic stability of systems (3.1) and (3.4) becomes equivalent.

Remark 3.4 The uncertain time-delay terms $h(\cdot)$ in system (3.4) are bounded by the product of the system state's absolute value and a nonlinear function with unknown parameters, as seen in Hua et al. (2005), Hua et al. (2008b), and Zhang et al. (2016). Time-delay is common in practical systems and poses a significant challenge to feedback control analysis and design. Here, only the boundedness of the delay d is required; no further information is necessary.

3.3 Adaptive Control Fully Actuated Nonlinear Systems with Uncertain Parameters

In this section, we consider control design for fully actuated nonlinear systems with uncertain parameters. An adaptive controller is designed and stability analysis of the resulting closed-loop system is then given based on the fully actuated system theory. Consider the following fully actuated nonlinear systems with uncertain parameters,

$$z^{(n)} = u + h\left(z^{(0 \sim n-1)}, \theta\right), \tag{3.5}$$

where variables and functions are defined in (3.4). Since $h(\cdot)$ contains no time-delay, it satisfies the following assumption.

Assumption 3.1 The uncertain nonlinear function $h(\cdot)$ satisfies

$$\|h(\cdot)\| \le \Theta \sum_{i=0}^{n-1} \left(\lambda_i\left(z^{(i)}\right)\left|z^{(i)}\right|\right), \tag{3.6}$$

where $\Theta = \Psi(\theta) > 0$ is a positive unknown parameter, and $\lambda_i(\cdot)$ is a known continuous function.

Following the fully actuated system theory, the adaptive controller can be designed as

$$u = -KZ + u^*, \tag{3.7}$$

where

$$Z = \left[z, \dot{z}, \ldots, z^{(n-1)} \right]^T \in \mathbb{R}^n,$$

$$K = [k_0, k_1, \ldots, k_{n-1}] \in \mathbb{R}^{1 \times n},$$

with the positive design parameters k_i, $i = 0, 1, \ldots, n - 1$. The term u^* is specifically designed to counteract the effects of the uncertain nonlinear function $h(\cdot)$. From (3.5) and (3.7), it can be verified that

$$\dot{Z} = \tilde{A}Z + L\left(h(\cdot) + u\right) = AZ + L\left(h(\cdot) + u^*\right), \tag{3.8}$$

where $L = [0, 0, \ldots, 0, 1]^T \in \mathbb{R}^n$, $A = \tilde{A} - LK$, and

$$\tilde{A} = \begin{bmatrix} 0 & & & \\ 0 & I_{(n-1)\times(n-1)} & \\ \vdots & & \\ 0\ 0 & \cdots & 0 \end{bmatrix}, \quad A = \begin{bmatrix} 0 & & & \\ 0 & I_{(n-1)\times(n-1)} & \\ \vdots & & \\ -k_0\ -k_1 & \cdots & -k_{n-1} \end{bmatrix}. \tag{3.9}$$

There is a positive definite symmetric matrix P, such that $A^T P + PA \leq -I$, where I is an identity matrix. Therefore, we have the following result.

Theorem 3.1 *For the fully actuated nonlinear systems (3.5) with uncertain parameters, under the Assumption 3.1, adaptive controller (3.7) with*

$$u^* = -4\hat{\Theta} L^T P Z \sum_{i=0}^{n-1} \lambda_i^2, \quad \dot{\hat{\Theta}} = 4Z^T P L L^T P Z \sum_{i=0}^{n-1} \lambda_i^2, \tag{3.10}$$

where $\hat{\Theta} = \Theta^2 - \tilde{\Theta}$ is the estimation of Θ^2, $\tilde{\Theta}$ is the estimation error, then the closed-loop system (3.5) is asymptotically stable.

Proof The candidate Lyapunov function is selected as

$$V = Z^T P Z + \frac{1}{2}\tilde{\Theta}^2. \tag{3.11}$$

According to (3.7) and (3.8), the derivative of the Lyapunov function V is given by

$$\dot{V} = Z^T \left(PA + A^T P \right) Z + 2Z^T PL \left(h(\cdot) + u^* \right) - \tilde{\Theta}\dot{\hat{\Theta}}. \tag{3.12}$$

Using the Young's inequality, it follows that

$$2Z^T PLh(\cdot) \le 4\Theta^2 Z^T PLL^T PZ \sum_{i=0}^{n-1} \lambda_i^2 + \frac{1}{2}Z^T Z, \tag{3.13}$$

where $\lambda_i(\cdot)$ is defined in Assumption 3.1.

Substituting (3.10) and (3.13) into (3.12), we can obtain that

$$\dot{V}_1 \le -\frac{1}{2}Z^T Z. \tag{3.14}$$

According to (3.14) and Barbalat's lemma, it can be concluded that

$$\lim_{t \to +\infty} Z(t) = 0, \tag{3.15}$$

which shows that Z asymptotically converges to the origin. Further, we know that system (3.5) is asymptotically stable. This completes the proof.

3.4   Adaptive Control Fully Actuated Nonlinear Systems with Uncertain Parameters and Time-Delays

In this section, we consider the control design for fully actuated nonlinear system (3.4) with both uncertain parameters and time delay. An adaptive straightforward and continuous controller is designed based on fully actuated system theory and stability analysis of the resulting closed-loop system is given based on the Lyapunov stability theorem.

To proceed, we introduce the following assumption regarding the nonlinear function $h(\cdot)$ in system (3.4).

Assumption 3.2 The uncertain nonlinear function $h(\cdot)$ satisfies the inequality

$$\|h(\cdot)\| \le \Theta \sum_{i=0}^{n-1} \left(\lambda_i \left(z^{(i)} \right) \left| z^{(i)} \right| + \sum_{j=1}^{n-i} \gamma_{ij} \left(z^{(i)}(t - jd) \right) \left| z^{(i)}(t - jd) \right| \right), \tag{3.16}$$

where $\Theta = \Psi(\theta) > 0$ is a positive unknown parameter, $\lambda_i(\cdot)$, $\gamma_{ij}(\cdot)$ are known, positive, continuous, and strictly increasing functions.

The λ_i and γ_{ij} are monotonically increasing functions with respect to their independent variable, there exists a positive continuous non-decreasing function

$\Phi(\cdot)$ such that

$$\sum_{i=0}^{n-1}\left(\lambda_i^2(z^{(i)}) + \sum_{j=1}^{n-i}\gamma_{ij}^2(z^{(i)})\right)(z^{(i)})^2 \leq \rho\Phi(Z^T P Z)Z^T P Z, \qquad (3.17)$$

where $\rho < \varepsilon$ is a positive constant.

Therefore, the adaptive continuous control input can be designed as follows:

$$u = -KZ - \frac{1}{2}\hat{\Theta}L^T P Z\Phi(W), \qquad (3.18)$$

$$\dot{\hat{\Theta}} = \left(\Phi(W)Z^T P L\right)^2, \qquad (3.19)$$

where $\Phi(\cdot)$ is defined in (3.17), $\hat{\Theta} = \Theta^2 - \tilde{\Theta}$ is the estimation of Θ^2, $\tilde{\Theta}$ is the estimation error, $W = Z^T P Z$. Then the following theorem can be obtained.

Theorem 3.2 *By designing the controller (3.18) and adaptive law (3.19), the closed-loop system composed of a fully actuated nonlinear system (3.4) achieves asymptotic stability when satisfying the given Assumption 3.2.*

Proof The candidate Lyapunov function is selected as

$$V = V_1 + V_2,$$

$$V_1 = \int_0^W \Phi(s)\, ds + \frac{1}{2}\tilde{\Theta}^2,$$

$$V_2 = \sum_{i=0}^{n-1}\sum_{j=1}^{n-i}\int_{t-jd}^t \gamma_{ij}^2\left(z^{(i)}(s)\right)\left(z^{(i)}(s)\right)^2 ds. \qquad (3.20)$$

According to (3.4), (3.18) and (3.20), the derivative of the Lyapunov function V_1 is given by

$$\dot{V}_1 = \Phi(W)\left(Z^T(PA + A^T P)Z + 2Z^T P L(h(\cdot) + u^*)\right) - \tilde{\Theta}\dot{\hat{\Theta}}. \qquad (3.21)$$

Using the Young's inequality, it follows that

$$2\Phi(W)Z^T P L h(\cdot) \leq \Theta^2\Phi^2(W)Z^T P L L^T P Z$$
$$+ \sum_{i=0}^{n-1}\left(\lambda_i^2(\cdot)(z^{(i)})^2 + \sum_{j=1}^{n-i}\gamma_{ij}^2(z^{(i)}(t-jd))(z^{(i)}(t-jd))^2\right)$$
$$\leq \Theta(Z^T P L\Phi(W))^2$$
$$+ \sum_{i=0}^{n-1}\left(\lambda_i^2(\cdot)(z^{(i)})^2 + \sum_{j=1}^{n-i}\gamma_{ij}^2(z^{(i)}(t-jd))(z^{(i)}(t-jd))^2\right), \qquad (3.22)$$

where $\lambda_i(\cdot)$ and $\gamma_{ij}(\cdot)$ are as defined in Assumption 3.2.

Substituting (3.18), (3.19) and (3.22) into (3.21), we can infer that

$$\dot{V}_1 \le -\varepsilon \Phi(W) Z^T P Z - \tilde{\Theta}(\tau - \dot{\hat{\Theta}})$$
$$+ \sum_{i=0}^{n-1} \left(\lambda_i^2(\cdot)(z^{(i)})^2 + \sum_{j=1}^{n-i} \gamma_{ij}^2 (z^{(i)}(t-jd))(z^{(i)}(t-jd))^2 \right), \qquad (3.23)$$

where $\tau = (Z^T P L \Phi(W))^2$, ε is a positive constant satisfying $A^T P + P A = -\varepsilon P$. From (3.20), the derivative of the Lyapunov function V_2 is expressed as

$$\dot{V}_2 = \sum_{i=0}^{n-1} \sum_{j=1}^{n-i} \left(\gamma_{ij}^2(\cdot)(z^{(i)})^2 - \gamma_{ij}^2 (z^{(i)}(t-jd))(z^{(i)}(t-jd))^2 \right). \qquad (3.24)$$

It can be further obtained from (3.17), (3.23), and (3.24) that

$$\dot{V} = \dot{V}_1 + \dot{V}_2 \le -c\Phi(W) Z^T P Z, \qquad (3.25)$$

where $c = \varepsilon - \rho > 0$ is a constant.

Finally, from (3.24) and Barbalat's lemma, it can be concluded that

$$\lim_{t \to +\infty} Z(t) = 0. \qquad (3.26)$$

which shows that Z asymptotically converges to the origin. Further, we know that system (3.4) is asymptotically stable. This completes the proof.

Remark 3.5 Since $\lambda_i(\cdot)$ and $\gamma_{ij}(\cdot)$ are known positive, continuous, and strictly increasing functions, there always exists a positive, nondecreasing, continuous function $\Phi(\cdot)$ such that inequality (3.17) holds. Specifically,

$$\sum_{i=0}^{n-1} \left(\lambda_i^2(z^{(i)}) + \sum_{j=1}^{n-i} \gamma_{ij}^2(z^{(i)}) \right)(z^{(i)})^2 \le \frac{Z^T P Z}{\lambda_{\min}(P)} \sum_{i=0}^{n-1} \left(\lambda_i^2(z^{(i)}) + \sum_{j=1}^{n-i} \gamma_{ij}^2(z^{(i)}) \right). \qquad (3.27)$$

Thus, we can choose

$$\Phi(Z^T P Z) \ge \frac{1}{\rho \lambda_{\min}(P)} \sum_{i=0}^{n-1} \left(\lambda_i^2(z^{(i)}) + \sum_{j=1}^{n-i} \gamma_{ij}^2(z^{(i)}) \right), \qquad (3.28)$$

ensuring the satisfaction of inequality (3.17). Since $\lambda_i(\cdot)$ and $\gamma_{ij}(\cdot)$ are strictly increasing functions, the existence of such a nondecreasing function $\Phi(\cdot)$ to satisfy (3.17) is guaranteed. With the chosen $\Phi(\cdot)$, inequality (3.28) holds. It is worth noting that the strict monotonicity of $\lambda_i(\cdot)$ and $\gamma_{ij}(\cdot)$ is a sufficient but not necessary condition for (3.17). In other words, as long as (3.28) is satisfied, a controller can be designed to achieve the control objective.

Remark 3.6 The selection of control parameters follows these steps: (1) Begin by selecting an appropriate parameter vector K and a positive definite symmetric matrix P, ensuring they satisfy the conditions specified in (3.9). (2) Choose the value of ε based on the requirements outlined in (3.9) to ensure system stability. (3) Determine ρ and $\Phi(\cdot)$ by solving inequality (3.17), ensuring that the chosen function $\Phi(\cdot)$ satisfies the necessary conditions. (4) Using the derived values of K, P, ε, ρ, and $\Phi(\cdot)$, finalize the adaptive law as given in (3.19) and construct the controller in its final form as shown in (3.18). This systematic approach ensures that the control parameters are appropriately selected to achieve the desired control objectives.

3.5   Simulation Results for Adaptive Controllers

In this section, to demonstrate the effectiveness of the proposed adaptive controller, examples of a standard fully actuated nonlinear system with uncertain parameters and a strict-feedback nonlinear time-delay system are presented.

Example 3.1 Consider the following standard fully actuated nonlinear system with uncertain parameters:

$$\dot{x}_1 = x_2,$$
$$\dot{x}_2 = u + h(x_1, x_2, \theta), \tag{3.29}$$

Define $z = x_1, \dot{z} = x_2$, then we have:

$$\ddot{z} = u + h(z, \dot{z}, \theta), \tag{3.30}$$

with

$$Z = [z, \dot{z}]^T \tag{3.31}$$

and the nonlinear term:

$$h(z, \dot{z}, \theta) = \theta_1 z^2(t) + \theta_2 \dot{z}^2$$
$$\leq \theta_1 |z||z| + \theta_2 |\dot{z}||\dot{z}|, \tag{3.32}$$

where

$$\Theta = \max\{\theta_1, \theta_2\},$$
$$\lambda_0(z(t)) = |z(t)|, \quad \lambda_1(\dot{z}(t)) = |\dot{z}(t)| \tag{3.33}$$

with $\theta_1 = \theta_2 = 0.8$ as unknown parameters.

Following the controller design process, the positive definite symmetric matrix P and parameter vector K are chosen as:

$$P = \begin{bmatrix} 1 & 1 \\ 1 & 2 \end{bmatrix}, \quad K = [2, 2]. \tag{3.34}$$

It follows that:

$$A^T P + PA \leq -P. \tag{3.35}$$

The final controller is then designed as:

$$u = -KZ - 4\hat{\Theta}L^T PZ(\lambda_0^2 + \lambda_1^2), \tag{3.36}$$
$$\dot{\hat{\Theta}} = 4Z^T PLL^T PZ(\lambda_0^2 + \lambda_1^2), \tag{3.37}$$

where

$$L = [0, 1]^T. \tag{3.38}$$

For simulation, the initial conditions are set as $z(0) = 1$, $\dot{z}(0) = -1$, and $\hat{\Theta}(0) = 0$. Simulation results (Figs. 3.1 and 3.2) show that the system state $Z(t)$ and control input $u(t)$ converge asymptotically to the origin, while the adaptive parameter $\hat{\Theta}$ remains bounded.

Example 3.2   Consider the strict-feedback nonlinear time-delay system:

$$\dot{x}_1 = x_2 + x_1(t - d),$$
$$\dot{x}_2 = u + \theta_1 x_1^2 + \theta_2 x_2(t - d). \tag{3.39}$$

Using the state transition

$$x_1 = z, \quad x_2 = \dot{z} - z(t - d), \tag{3.40}$$

the system can be transformed into

$$\ddot{z} = u + (\theta_2 + 1)\dot{z}(t - d) + \theta_1 z^2 - \theta_2 z(t - 2d), \tag{3.41}$$

Fig. 3.1 The responses of the system states $z(t)$ and $\dot{z}(t)$. (**a**) The response of z. (**b**) The response of $\dot{z}$

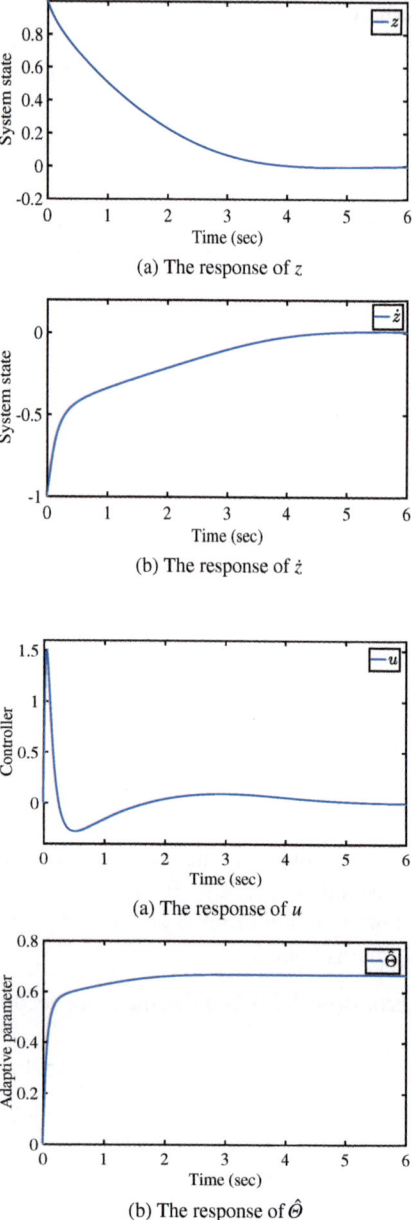

(a) The response of z

(b) The response of $\dot{z}$

Fig. 3.2 The responses of the controller $u(t)$ and adaptive parameter $\hat{\Theta}(t)$. (**a**) The response of u. (**b**) The response of $\hat{\Theta}$

(a) The response of u

(b) The response of $\hat{\Theta}$

where

$$Z = [z, \dot{z}]^T, \tag{3.42}$$

and $d = 0.5$, and the nonlinear term $h(\cdot)$ is given by

$$h(\cdot) = (\theta_2 + 1)\dot{z}(t - d) + \theta_1 z^2 - \theta_2 z(t - 2d)$$
$$\leq \Theta \big(|z||z| + |z(t - 2d)| + |\dot{z}(t - d)|\big). \tag{3.43}$$

From the above, we have

$$\Theta = \max\{\theta_1, \theta_2 + 1\} \quad \lambda_0(z) = |z(t)|, \quad \lambda_1(\dot{z}) = 0,$$
$$\gamma_{01}(z(t - d)) = 0, \quad \gamma_{02}(z(t - 2d)) = 1, \quad \gamma_{11}(\dot{z}(t - d)) = 1, \tag{3.44}$$

where $\theta_1 = 0.8$ and $\theta_2 = 0.6$ are unknown parameters.

Using the proposed method, we design the controller as follows. By choosing $\varepsilon = 3$, we select

$$P = \begin{bmatrix} 6 & 3 \\ 3 & 2 \end{bmatrix}, \quad K = \begin{bmatrix} 3 \\ 3 \end{bmatrix}^T. \tag{3.45}$$

From (3.43) and (3.45), with $\rho = 0.3$, we define

$$\Phi(Z^T P Z) = 10 Z^T P Z + 6. \tag{3.46}$$

The final controller is

$$u = -KZ - \frac{1}{2}\hat{\Theta} L^T P Z \Phi(Z^T P Z), \quad \dot{\hat{\Theta}} = (\Phi(Z^T P Z)Z^T P L)^2, \tag{3.47}$$

where

$$L = [0, 1]^T. \tag{3.48}$$

The initial values are set as $V(s) = [-1, 1]^T$, and $\hat{\Theta}(0) = 0$. The simulation results are presented in Figs. 3.3 and 3.4, showing that the system state $x(t)$, $Z(t)$ and the control input $u(t)$ asymptotically converge to the origin, while the adaptive parameter $\hat{\Theta}$ remains bounded.

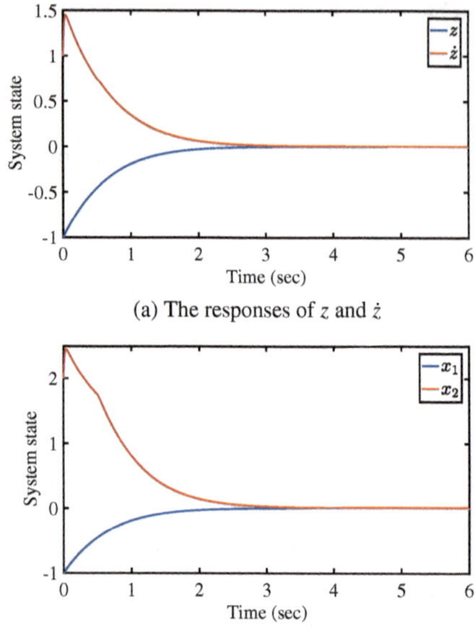

Fig. 3.3 The responses of the states $z(t)$, $\dot{z}(t)$, $x_1(t)$ and $x_2(t)$. (**a**) The responses of z and $\dot{z}$. (**b**) The responses of states x_1, x_2

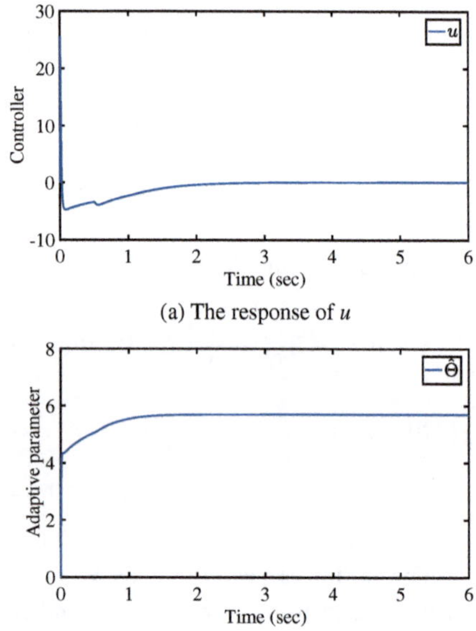

(a) The responses of z and $\dot{z}$

(b) The responses of states x_1, x_2

Fig. 3.4 The responses of the controller $u(t)$, and adaptive parameter $\hat{\Theta}$. (**a**) The response of u. (**b**) The response of $\hat{\Theta}$

(a) The response of u

(b) The response of $\hat{\Theta}$

3.6   Notes and References

In this chapter, we address the global adaptive control problem for nonlinear systems with uncertain parameters and time delay. By utilizing the fully-actuated systems approach, two adaptive feedback controllers are designed to stabilize fully-actuated nonlinear systems with uncertain parameters and strict-feedback nonlinear systems with both uncertain parameters and time delays, respectively. The nonlinear time-delay and uncertainties are bounded by the product of the system state's absolute value and a nonlinear function containing unknown parameters. It is shown that the states of the resulting closed-loop systems converge to the origin asymptotically. Simulation examples are provided to illustrate the effectiveness of our results.

Chapter 4
Adaptive Input-Constrained Control of High-Order Fully Actuated Nonlinear Time-Delay Systems with Unmodeled Dynamics

Building upon the adaptive control strategies in the previous chapter, this chapter extends the discussion by considering the challenges posed by input dead zones and unmodeled dynamics in high-order fully actuated nonlinear systems with time delays. Input dead zones, such as actuator saturation or physical limitations; Unmodeled dynamics, such as friction or unmodeled higher-order effects, can significantly degrade control performance. To address these challenges, we develop adaptive input-constrained controllers that ensure stability and robustness. By combining HOFA system theory with adaptive control techniques, we employ Lyapunov-Krasovskii functional methods to guarantee system stability and convergence while respecting input constraints.

4.1 Overview

Building upon the systematic exposition in Chap. 3 regarding the developmental trajectory, fundamental framework, and representative applications of HOFA systems theory in robust control (Duan 2021c), adaptive control (Duan 2021d), hybrid robust-adaptive control (Duan 2021e), disturbance attenuation (Duan 2021f), and decoupling (Wang et al. 2022c), this chapter focuses on its latest theoretical extensions. As outlined previously, the HOFA methodology provides systematic solutions for analyzing and controlling high-order nonlinear systems by constructing linear time-invariant closed-loop systems with arbitrarily assignable eigenstructures. Recent advancements have extended this approach to discrete-time systems with time-varying delays in Liu (2022a) and continuous-time delay systems through adaptive control synthesis in Ning et al. (2022b). However, significant theoretical challenges remain in tackling complex nonlinear phenomena that involve coupled time delays, unmodeled dynamics, and dead-zone nonlinearities, necessitating further research to improve methodological universality.

Time delays, ubiquitous in practical systems from power networks to chemical reactors (Zhang et al. 2013; Tong et al. 2011), often interact with unmodeled dynamics to trigger instability or performance deterioration. Current stabilization paradigms predominantly employ Lyapunov-Krasovskii functional methods and Lyapunov-Razumikhin techniques (Zhang et al. 2018a). Noteworthy contributions include Zhang et al.'s memoryless state feedback design via tailored Lyapunov-Krasovskii functionals (Zhang & Raissi 2021), and nonlinear event-triggered controllers derived from novel Lyapunov-Razumikhin function families (Hua et al. 2009; Jin et al. 2010; Zhao & Lin 2021; Khattak & Iqbal 2006). For systems with unmodeled dynamics, Hua et al. (2017a) developed adaptive fuzzy prescribed performance control for switched time-delay systems using supply-rate modification, while dynamic gain techniques addressed global stability in cascaded systems with unknown delays in (Zhang et al. 2018a, 2017b). Despite these advances, HOFA-based approaches for coupled time-delay/unmodeled dynamics systems remain conspicuously absent in literature.

Dead-zone nonlinearities, prevalent in hydraulic actuators, gear transmissions, and servo mechanisms, necessitate specialized compensation strategies. Current methodologies include adaptive state feedback designs via dead-zone decomposition into quasi-linear and disturbance-like components (Hua et al. 2008b; Zhang and Ge 2009, with Hua et al. 2017b) advancing adaptive estimation of asymmetric dead-zone slope bounds. Lu et al. (2019) further proposed adaptive inverse controllers for non-parametric dead-zone compensation, demonstrating effective dynamic characteristic mitigation.

4.2 Problem Formulation

This chapter addresses the control problem for nonlinear HOFA systems with time delays, unmodeled dynamics, and unknown dead-zone inputs. The central aim of this study is to develop an adaptive controller using the HOFA system framework.

4.2.1 Dead Zone Model

In this section, we consider the following dead-zone model:

$$u(t) = D(v(t)) = \begin{cases} m_r(v(t) - b_r), & v(t) \geq b_r \\ 0, & b_l < v(t) < b_r \\ m_l(v(t) - b_l), & v(t) \leq b_l \end{cases} \tag{4.1}$$

let $u(t) \in \mathbb{R}$ denote the dead-zone output, $D(\cdot) \in \mathbb{R}$ represent the asymmetric dead-zone nonlinear characteristic, and $v(t) \in \mathbb{R}$ is the dead-zone input. Here,

m_l and m_r correspond to the left and right slopes of the dead-zone characteristic, respectively, while b_l and b_r define the left and right breakpoints of the input nonlinearity. The dead-zone width is characterized by $b_r - b_l$.

The actual actuator $u(t)$ can be transformed as follows:

$$u(t) = \Upsilon^T \varpi, \tag{4.2}$$

where

$$\Upsilon = [m_r, m_r b_r, m_l, m_l b_l]^T \in \mathbb{R}^4,$$

$$\varpi = [vc_r, -c_r, vc_l, -c_l]^T \in \mathbb{R}^4,$$

and

$$c_r = \begin{cases} 1, & v(t) \geq b_r \\ 0, & else \end{cases}, \quad c_l = \begin{cases} 1, & v(t) \leq b_l \\ 0, & else \end{cases}. \tag{4.3}$$

To mitigate the dead-zone nonlinearity, an adaptive dead-zone inverse compensator is formulated:

$$v(t) = \widehat{DI}(u_d(t)) = \frac{u_d(t) + \widehat{m_r b_r}}{\widehat{m_r}} \beta_r + \frac{u_d(t) + \widehat{m_l b_l}}{\widehat{m_l}} \beta_l, \tag{4.4}$$

where $\widehat{m_r b_r}, \widehat{m_r}, \widehat{m_l b_l}$ and $\widehat{m_l}$ are the estimates of $m_r b_r, m_r, m_l b_l$ and m_l, respectively. $u_d(t)$ is the input of the dead-zone inverse. Besides, $\hbar$ is a positive constant, the bounded functions β_r and β_l are as follows:

$$\beta_l = \frac{\pi - 2\arctan(\hbar u_d(t))}{2\pi}, \beta_r = \frac{\pi + 2\arctan(\hbar u_d(t))}{2\pi}. \tag{4.5}$$

The adaptive dead-zone inverse can also be converted into the following form:

$$u_d(t) = \widehat{\Upsilon}^T \varpi_d, \tag{4.6}$$

where

$$\widehat{\Upsilon} = \left[\widehat{m_r}, \widehat{m_r b_r}, \widehat{m_l}, \widehat{m_l b_l}\right]^T = [\widehat{\Upsilon}_1, \widehat{\Upsilon}_2, \widehat{\Upsilon}_3, \widehat{\Upsilon}_4]^T \in R^4,$$

$$\varpi_d = [v\beta_r, -\beta_r, v\beta_l, -\beta_l]^T \in R^4.$$

The compensation error between the dead-zone output $u(t)$ and the designed controller $u_d(t)$ is defined as

$$u(t) - u_d(t) = \Upsilon^T \varpi - \widehat{\Upsilon}^T \varpi_d = \widetilde{\Upsilon}^T \varpi_d + \rho(t), \tag{4.7}$$

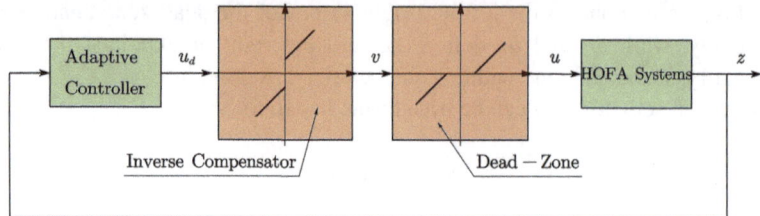

Fig. 4.1 Block diagram of the closed-loop system

with

$$\rho(t) = \Upsilon^T(\varpi - \varpi_d) = m_r(v - b_r)(c_r - \beta_r) + m_l(v - b_l)(c_l - \beta_l), \quad (4.8)$$

where $|\rho(t)| \leq \rho$, ρ is a positive constant, and $\widetilde{\Upsilon} = \Upsilon - \widehat{\Upsilon}$ is the estimation error.

Remark 4.1 A residual compensation error $\rho(t)$ inevitably occurs when using the inverse dead-zone for compensation. However, due to the boundedness of $c_r - \beta r$ and $cl - \beta_l$, $\rho(t)$ remains bounded. Thus, $\rho(t)$ can be reduced by increasing $\hbar$ in (4.5). Additionally, beyond increasing $\hbar$, robust control is employed in the adaptive controller design to further reduce the error. The dead-zone and inverse compensator interplay is illustrated in Fig. 4.1's block diagram.

4.2.2 Nonlinear Time-Delay Systems with Unmodeled Dynamics

In this section, we address the control challenge for a class of time-delay nonlinear HOFA systems incorporating unmodeled dynamics and an unidentified dead-zone input:

$$\dot{E} = f(E, E_d, z, z_d, \theta),$$
$$z^{(n)} = h\left(E, E_d, z^{(0 \sim n-1)}, \cdots, z^{(0)}, z_d^{(0 \sim n-1)}, \cdots, z_d^{(0)}, \theta\right) + u(t), \quad (4.9)$$

where $z \in \mathbb{R}$ is the measurable state, and

$$E = [e_0, e_1, \cdots, e_{n-1}]^T \in \mathbb{R}^n$$

represents the unmodeled dynamic states, $\theta \in \mathbb{R}^s$ denotes an unknown parameter vector, $f(\cdot) \in \mathbb{R}^n$ is a continuously differentiable vector field, $h(\cdot) \in \mathbb{R}$ corresponds to an uncertain nonlinear function, $(\circ)_d$ is introduced to denote the delayed term $(\circ)(t - d(t))$, $d(t) \geq 0$ denotes the system's unknown time-varying delay, constrained by $d\,(t) \leq d^*$ with its rate satisfying $\dot{d}\,(t) \leq \bar{d} < 1$, and $u\,(t) \in \mathbb{R}$ signifies the dead-zone output signal.

To develop an adaptive control strategy addressing the time-delay nonlinear HOFA system (4.9) with unmodeled dynamics and unidentified dead-zone characteristics, we impose the following assumptions on system (4.9):

Assumption 4.1 The uncertain nonlinear function $h(\cdot)$ satisfies

$$
|h(\cdot)| \leq R\,(\theta) \left[\Delta_1(E)\,\|E\| + \Delta_2(E_d)\,\|E_d\| + \sum_{i=0}^{n-1} \gamma_i\left(z^{(i)}\right) \left| z^{(i)} \right| \right.
$$
$$
\left. + \sum_{i=0}^{n-1} \lambda_i\left(z_d^{(i)}\right) \left| z_d^{(i)} \right| \right], \tag{4.10}
$$

where $R\,(\theta)$ is a positive smooth function, $\Delta_1(\cdot)$, $\Delta_2(\cdot)$, $\gamma_i\,(\cdot)$ and $\lambda_i\,(\cdot)$ are known continuous positive increasing functions.

Assumption 4.2 There is a $\mathbb{C}^1$ Lyapunov function $U\,(E)$, which is positive definite and satisfies

$$
\underline{\alpha}\,(E) \leq U\,(E) \leq \overline{\alpha}\,(E), \tag{4.11}
$$

$$
\dot{U}\,(E) = \dot{E}^T \frac{\partial U\,(E)}{\partial E} \leq -aU\,(E) + \varphi_1\,(|z|) + \varphi_2\,(|z_d|), \tag{4.12}
$$

where a is a positve constant, $\underline{\alpha}\,(\cdot)$, $\overline{\alpha}\,(\cdot)$, $\varphi_1\,(\cdot)$ and $\varphi_2\,(\cdot)$ are class K_∞ functions.

Assumption 4.3 The function φ_1, φ_2, Δ_1 and Δ_2 satisfy the following conditions:

$$
\lim_{t \to 0^+} \sup \frac{\Delta_1^2\,(t)\,\|t\|^2}{U\,(t)} < +\infty, \quad \lim_{t \to 0^+} \sup \frac{\varphi_1\,(t)}{t^2} < +\infty, \tag{4.13}
$$

$$
\lim_{t \to 0^+} \sup \frac{\Delta_2^2\,(t)\,\|t\|^2}{U\,(t)} < +\infty, \quad \lim_{t \to 0^+} \sup \frac{\varphi_2\,(t)}{t^2} < +\infty. \tag{4.14}
$$

Assumption 4.4 The parameters $m_r > 0$, $m_l > 0$, $b_r > 0$ and $b_l < 0$ are unknown constants.

Remark 4.2 Assumption 4.1 indicates that the smooth function $h(\cdot)$, involving unmodeled dynamics E and uncertain parameters θ is upper-bounded by the product of system states and the differentiable function $R\,(\theta)$. Furthermore, Assumptions 4.2 and 4.3 align with standard requirements for unmodeled dynamic subsystems as discussed in Hua et al. (2017a), verifying that these dynamics adhere to the input-to-state stability (ISS) condition. By leveraging Assumption 4.3, the terms $\Delta_1(\cdot)$ and $\Delta_2(\cdot)$ appearing in (4.47) can be systematically handled.

4.3 Adaptive Input-Constrained Controller Design

Initially, a HOFA-based adaptive control framework is developed to achieve stabilization of system (4.9).

Given the fulfillment of Assumptions 4.1–4.4 for system (4.9), the adaptive controller architecture is constructed as follows:

$$v(t) = \frac{u_d(t) + \widehat{\Upsilon}_2}{\widehat{\Upsilon}_1} \beta_r(u_d(t)) + \frac{u_d(t) + \widehat{\Upsilon}_4}{\widehat{\Upsilon}_3} \beta_l(u_d(t)),$$

$$u_d(t) = -A^{0\sim m-1} z^{(0\sim n-1)} - u^*(t),$$

$$u^*(t) = \frac{1}{2}\widehat{\Theta} Q(W) P_L^T z^{(0\sim n-1)} + \frac{\rho^2}{2\kappa} Q(W) P_L^T z^{(0\sim n-1)}, \qquad (4.15)$$

with the adaptive law

$$\dot{\widehat{\Theta}} = [\left(z^{(0\sim n-1)}\right)^T P_L Q(W)]^2 - \widehat{\Theta},$$

$$\dot{\widehat{\Upsilon}} = 2Q(W)\left(z^{(0\sim n-1)}\right)^T P_L \Gamma \varpi_d - \Gamma \widehat{\Upsilon}, \qquad (4.16)$$

where $\widehat{\Theta} = \Theta - \widetilde{\Theta}$ defines the estimated value of $\Theta = R(\theta)$, with $\widetilde{\Theta}$ denoting the estimation error, and $\kappa > 0$ serves as a tuning parameter. Furthermore, $A_i \in \mathbb{R}^{r\times r}$ $(i = 1, 2, \ldots, n-1)$ are predefined matrices complying with Lemma 2.5, while P represents a symmetric positive definite matrix fulfilling the requirements specified in Lemma 2.6, where

$$z^{(0\sim n-1)} = \left[z, \dot{z}, \cdots, z^{(n-1)}\right]^T \in \mathbb{R}^n,$$

$$W = \left(z^{(0\sim n-1)}\right)^T P\left(z^{(0\sim n-1)}\right).$$

The positive nondecreasing continuous function $Q(\cdot) \in R \geq 0$ satisfies

$$\sum_{i=0}^{n-1}\left[\gamma_i^2\left(z^{(i)}\right) + \frac{e^{\varepsilon d^*}}{1-d}\lambda_i^2\left(z^{(i)}\right)\right]\left(z^{(i)}\right)^2 \leq \rho Q(W) W, \qquad (4.17)$$

where ρ is a positive constant. Then, the solution of the time-delay closed-loop systems is semi-globally bounded.

Proof Firstly, substituting $u_d = -A^{0 \sim m-1} z^{(0 \sim n-1)} - u^*$ into the time-delay nonlinear HOFA systems (4.9), one can easily obtain the closed-loop system as

$$z^{(n)} + A^{0 \sim m-1} z^{(0 \sim n-1)} = h\left(E, E_d, z^{(0 \sim n-1)}, \cdots, z^{(0)}, z_d^{(0 \sim n-1)}, \cdots, z_d^{(0)}, \theta\right)$$
$$+ \widetilde{\varUpsilon}^T \varpi_d + \rho(t) - u^*.$$
(4.18)

Then, we rewrite the above HOFA systems (4.18) in the following state-space form:

$$\dot{z}^{(0 \sim n-1)} = \varPhi\left(A^{0 \sim m-1}\right) z^{(0 \sim n-1)} + \begin{bmatrix} 0 \\ h(\cdot) + \widetilde{\varUpsilon}^T \varpi_d + \rho(t) - u^* \end{bmatrix}.$$
(4.19)

4.4 Lyapunov-Based Stability Analysis for Nonlinear Time-Delay Systems

To validate the stability of the proposed controller, the following Lyapunov function is selected for system (4.9):

$$V = V_0 + V_1 + V_2,$$
(4.20)

with

$$V_0 = \int_0^{U(E)} q(s) \mathrm{d}s,$$
(4.21)

$$V_1 = \int_0^W Q(s) \mathrm{d}s + \frac{1}{2} \widetilde{\varTheta}^2 + \frac{1}{2} \widetilde{\varUpsilon}^T \varGamma^{-1} \widetilde{\varUpsilon},$$
(4.22)

$$V_2 = \frac{e^{\varepsilon d^*}}{1 - d} \left[\int_{t-d(t)}^t e^{-\varepsilon(t-s)} \Delta_2^2(E(s)) \| E(s) \|^2 \mathrm{d}s \right.$$
$$+ \sum_{i=0}^{n-1} \int_{t-d(t)}^t e^{-\varepsilon(t-s)} \lambda_i^2(z(s)^{(i)})(z(s)^{(i)})^2 \mathrm{d}s$$
$$\left. + \int_{t-d(t)}^t e^{-\varepsilon(t-s)} \widehat{\varphi}_2(|z(s)|) \mathrm{d}s \right],$$
(4.23)

where $q(\cdot) : [0, \infty) \to [1, \infty)$ is a $\mathbb{C}^0$ nondecreasing function and satisfies

$$\frac{a}{4} q [U(E)] U(E) \geq \left(\Delta_1^2(E) + \frac{e^{\varepsilon d^*}}{1-\bar{d}} \Delta_2^2(E) \right) \|E\|^2 .$$

Moreover, $\Gamma \in \mathbb{R}^{4 \times 4}$ constitutes a symmetric positive definite matrix.

First, we analyze the Lyapunov function V_0, and evidently, V_0 is $\mathbb{C}^1$-smooth and positive definite. Taking the time derivative of (4.21) yields:

$$\dot{V}_0 = q[U(E)][-aU(E) + \varphi_1(|z|) + \varphi_2(|z_d|)] . \tag{4.24}$$

Then, the discussion of (4.24) is divided into two cases:

Case I: If $-\frac{a}{2} U(E) + \varphi_1(|z|) + \varphi_2(|z_d|) \leq 0$, we can get that

$$\dot{V}_0 \leq -\frac{a}{2} q[U(E)] U(E) . \tag{4.25}$$

Case II: If $-\frac{a}{2} U(E) + \varphi_1(|z|) + \varphi_2(|z_d|) > 0$, we can obtain

$$U(E) < \frac{2}{a} (\varphi_1(|z|) + \varphi_2(|z_d|)) . \tag{4.26}$$

Taking the inverse function of (4.26), it becomes

$$E \leq U^{-1} \left(\frac{2}{a} (\varphi_1(|z|) + \varphi_2(|z_d|)) \right) . \tag{4.27}$$

Then, we have

$$q[U(E)] \leq q[\bar{\alpha}(E)] \leq q\left(\bar{\alpha} \left(U^{-1} \left(\frac{2}{a} (\varphi_1(|z|) + \varphi_2(|z_d|)) \right) \right) \right) . \tag{4.28}$$

In view of (4.24) and (4.28), we obtain

$$\dot{V}_0 \leq -\frac{a}{2} q[U(E)] U(E)$$
$$+ q\left(\bar{\alpha} \left(U^{-1} \left(\frac{2}{a} (\varphi_1(|z|) + \varphi_2(|z_d|)) \right) \right) \right) (\varphi_1(|z|) + \varphi_2(|z_d|)) . \tag{4.29}$$

Combining the two cases above, the derivative of V_0 can be estimated as

$$\dot{V}_0 \leq -\frac{a}{2} q[U(E)] U(E) + \widehat{\varphi}_1(|z|) + \widehat{\varphi}_2(|z_d|) . \tag{4.30}$$

In addition, due to $q[U(E)] \geq 1$, we should point out that the relationship between two continuous class K_∞ functions $\widehat{\varphi}_i(|z|)$ and $\varphi_i(|z|)$ is $\widehat{\varphi}_i(|z|) \geq \varphi_i(|z|), i = 1, 2$.

Then, the time derivative of V_1 is

$$\dot{V}_1 = Q(W)\left[\left(\dot{z}^{(0\sim n-1)}\right)^T P z^{(0\sim n-1)} + \left(z^{(0\sim n-1)}\right)^T P \dot{z}^{(0\sim n-1)}\right]$$
$$- \widetilde{\Theta}\dot{\widehat{\Theta}} - \widetilde{\Upsilon}^T \Gamma^{-1} \dot{\widehat{\Upsilon}}. \tag{4.31}$$

Substituting (4.19) into (4.31), we obtain

$$\dot{V}_1 = Q(W)\left[\left(z^{(0\sim n-1)}\right)^T (\Phi^T P + P\Phi) z^{(0\sim n-1)}\right.$$
$$\left. + 2\left(z^{(0\sim n-1)}\right)^T P \begin{pmatrix} 0 \\ h(\cdot) + \widetilde{\Upsilon}^T \varpi_d + \rho(t) - u^* \end{pmatrix}\right]$$
$$- \widetilde{\Theta}\dot{\widehat{\Theta}} - \widetilde{\Upsilon}^T \Gamma^{-1} \dot{\widehat{\Upsilon}}. \tag{4.32}$$

Moreover, combining Lemma 2.6, we can obtain

$$\dot{V}_1 \leq Q(W)\left[-\mu W + 2\left(z^{(0\sim n-1)}\right)^T P_L\left(h(\cdot) + \widetilde{\Upsilon}^T \varpi + d(t) - u^*\right)\right]$$
$$- \widetilde{\Theta}\dot{\widehat{\Theta}} - \widetilde{\Upsilon}^T \Gamma^{-1} \dot{\widehat{\Upsilon}}$$
$$= -\mu Q(W)W - \widetilde{\Theta}\dot{\widehat{\Theta}} - \widetilde{\Upsilon}^T \Gamma^{-1} \dot{\widehat{\Upsilon}} + 2Q(W)\left(z^{(0\sim n-1)}\right)^T P_L h(\cdot)$$
$$- 2Q(W)\left(z^{(0\sim n-1)}\right)^T P_L u^* + 2Q(W)\left(z^{(0\sim n-1)}\right)^T P_L \rho(t)$$
$$+ 2Q(W)\left(z^{(0\sim n-1)}\right)^T P_L \widetilde{\Upsilon}^T \varpi_d. \tag{4.33}$$

From Assumption 4.1 and Lemma 2.7, it is easy to get that

$$2Q(W)\left(z^{(0\sim n-1)}\right)^T P_L h(\cdot) \leq \Theta[\left(z^{(0\sim n-1)}\right)^T P_L Q(W)]^2 + \Delta_1^2(E)\|E\|^2$$
$$+ \Delta_2^2(E_d)\|E_d\|^2 + \sum_{i=0}^{n-1} \gamma_i^2\left(z^{(i)}\right)\left(z^{(i)}\right)^2 + \sum_{i=0}^{n-1} \lambda_i^2\left(z_d^{(i)}\right)\left(z_d^{(i)}\right)^2. \tag{4.34}$$

Based on the (4.33) and (4.34), one can easily prove that

$$
\begin{aligned}
\dot{V}_1 \leq & -\mu Q\left(W\right) W + \Theta\left[\left(z^{(0 \sim n-1)}\right)^T P_L Q\left(W\right)\right]^2 - 2Q\left(W\right)\left(z^{(0 \sim n-1)}\right)^T P_L u^* \\
& -\tilde{\Upsilon}^T \Gamma^{-1} \hat{\Upsilon} + 2Q(W)\left(z^{(0 \sim n-1)}\right)^T P_L \rho\left(t\right) + \Delta_1^2(E)\|E\|^2 \\
& +\Delta_2^2(E_d)\|E_d\|^2 - \tilde{\Theta}\hat{\Theta} + 2Q(W)\left(z^{(0 \sim n-1)}\right)^T P_L \tilde{\Upsilon}^T \varpi_d \\
& +\sum_{i=0}^{n-1} \gamma_i^2\left(z^{(i)}\right)\left(z^{(i)}\right)^2 + \sum_{i=0}^{n-1} \lambda_i^2\left(z_d^{(i)}\right)\left(z_d^{(i)}\right)^2 .
\end{aligned}
\tag{4.35}
$$

Moreover, in view (4.15), we have

$$
\begin{aligned}
-2Q(W)\left(z^{(0 \sim n-1)}\right)^T P_L u^* = & -\left[\hat{\Theta}\left[\left(z^{(0 \sim n-1)}\right)^T P_L Q(W)\right]^2\right. \\
& \left.+\frac{\rho^2}{2\kappa}\left[\left(z^{(0 \sim n-1)}\right)^T P_L Q(W)\right]^2\right] .
\end{aligned}
\tag{4.36}
$$

By combining (4.35) and (4.36), we obtain

$$
\begin{aligned}
\dot{V}_1 \leq & -\mu Q\left(W\right) W + \Theta\left[\left(z^{(0 \sim n-1)}\right)^T P_L Q\left(W\right)\right]^2 \\
& -\hat{\Theta}\left[\left(z^{(0 \sim n-1)}\right)^T P_L Q(W)\right]^2 \\
& -\frac{\rho^2}{2\kappa}\left[\left(z^{(0 \sim n-1)}\right)^T P_L Q(W)\right]^2 + 2Q(W)\left(z^{(0 \sim n-1)}\right)^T P_L \rho\left(t\right) \\
& +2Q(W)\left(z^{(0 \sim n-1)}\right)^T P_L \tilde{\Upsilon}^T \varpi_d + \sum_{i=0}^{n-1} \gamma_i^2\left(z^{(i)}\right)\left(z^{(i)}\right)^2 \\
& +\sum_{i=0}^{n-1} \lambda_i^2\left(z_d^{(i)}\right)\left(z_d^{(i)}\right)^2 \\
& +\Delta_1^2(E)\|E\|^2 + \Delta_2^2(E_d)\|E_d\|^2 - \tilde{\Upsilon}^T \Gamma^{-1} \hat{\Upsilon} - \tilde{\Theta}\hat{\Theta} .
\end{aligned}
\tag{4.37}
$$

Considering $|\rho(t)| \leq \rho$, we derive

$$-\frac{\rho^2}{4\kappa}\left[\left(z^{(0 \sim n-1)}\right)^T P_L Q(W)\right]^2 + Q(W)\left(z^{(0 \sim n-1)}\right)^T P_L \rho(t)$$

$$\leq -\frac{\rho^2}{4\kappa}\left[\left(z^{(0 \sim n-1)}\right)^T P_L Q(W)\right]^2 + \left|Q(W)\left(z^{(0 \sim n-1)}\right)^T P_L\right| |\rho(t)|$$

$$\leq -\frac{\rho^2}{4\kappa}\left[\left(z^{(0 \sim n-1)}\right)^T P_L Q(W)\right]^2 + \rho\left|Q(W)\left(z^{(0 \sim n-1)}\right)^T P_L\right|$$

$$\leq \kappa. \tag{4.38}$$

With the help of (4.16), one can obtain

$$\dot{V}_1 \leq -\mu Q(W)W + 2\kappa + \Delta_1^2(E)\|E\|^2 + \Delta_2^2(E_d)\|E_d\|^2$$

$$+ \sum_{i=0}^{n-1} \gamma_i^2(z^{(i)})\left(z^{(i)}\right)^2 - \tilde{\Theta}\hat{\Theta} + \sum_{i=0}^{n-1} \lambda_i^2(z_d^{(i)})\left(z_d^{(i)}\right)^2 - \tilde{\Upsilon}^T\hat{\Upsilon}. \tag{4.39}$$

From the Youngs inequality, we can know

$$\tilde{\Theta}\hat{\Theta} \leq \frac{1}{2}\Theta^2 - \frac{1}{2}\tilde{\Theta}^2, \tag{4.40}$$

$$\tilde{\Upsilon}^T\hat{\Upsilon} \leq \frac{1}{2}\Upsilon^T\Upsilon - \frac{\tilde{\Upsilon}^T\Gamma^{-1}\tilde{\Upsilon}}{2\lambda_{\max}\left(\Gamma^{-1}\right)}. \tag{4.41}$$

Substituting (4.40) and (4.41) into (4.39) implies that

$$\dot{V}_1 \leq -\mu Q(W)W + 2\kappa + \frac{1}{2}\Upsilon^T\Upsilon - \frac{\tilde{\Upsilon}^T\Gamma^{-1}\tilde{\Upsilon}}{2\lambda_{\max}\left(\Gamma^{-1}\right)} + \sum_{i=0}^{n-1}\gamma_i^2\left(z^{(i)}\right)\left(z^{(i)}\right)^2$$

$$+ \sum_{i=0}^{n-1}\lambda_i^2\left(z_d^{(i)}\right)\left(z_d^{(i)}\right)^2 + \Delta_1^2(E)\|E\|^2$$

$$+ \Delta_2^2(E_d)\|E_d\|^2 + \frac{1}{2}\Theta^2 - \frac{1}{2}\tilde{\Theta}^2. \tag{4.42}$$

From (4.23), the time derivative of Lyapunov function V_2 is shown as

$$\dot{V}_2 = -\varepsilon V_2 + \frac{e^{\varepsilon d^*}}{1-\bar{d}}\left[e^{-\varepsilon t}\left(e^{\varepsilon t}\Delta_2^2(E)\|E\|^2\right.\right.$$

$$\left.\left. - e^{\varepsilon(t-d(t))}\Delta_2^2(E_d)\|E_d\|^2\left(1 - \dot{d}(t)\right)\right)\right.$$

$$+ e^{-\varepsilon t} \sum_{i=0}^{n-1} \left(e^{\varepsilon t} \lambda_i^2 \left(z^{(i)} \right) \left(z^{(i)} \right)^2 - e^{\varepsilon (t - d(t))} \lambda_i^2 \left(z_d^{(i)} \right) \left(z_d^{(i)} \right)^2 \left(1 - \dot{d}(t) \right) \right)$$

$$+ e^{-\varepsilon t} \left(e^{\varepsilon t} \widehat{\varphi}_2 \left(|z| \right) - e^{\varepsilon (t - d(t))} \widehat{\varphi}_2 \left(|z_d| \right) \left(1 - \dot{d}(t) \right) \right) \Big]$$

$$\leq -\varepsilon V_2 + \frac{e^{\varepsilon d^*}}{1 - \bar{d}} \left[\Delta_2^2 (E) \, \|E\|^2 + \sum_{i=0}^{n-1} \lambda_i^2 \left(z^{(i)} \right) \left(z^{(i)} \right)^2 + \widehat{\varphi}_2 \left(|z| \right) \right]$$

$$- \sum_{i=0}^{n-1} \lambda_i^2 \left(z_d^{(i)} \right) \left(z_d^{(i)} \right)^2 - \widehat{\varphi}_2 \left(|z_d| \right) - \Delta_2^2 (E_d) \, \|E d\|^2 . \tag{4.43}$$

Combining the above analysis, we can obtain

$$\dot{V} = \dot{V}_0 + \dot{V}_1 + \dot{V}_2$$

$$\leq -\mu Q (W) \, W + 2\kappa + \frac{1}{2} \Theta^2 - \frac{1}{2} \widetilde{\Theta}^2 + \frac{1}{2} \Upsilon^T \Upsilon - \frac{\widetilde{\Upsilon}^T \Gamma^{-1} \widetilde{\Upsilon}}{2 \lambda_{\max} \left(\Gamma^{-1} \right)} + \Delta_1^2 (E) \, \|E\|^2$$

$$+ \sum_{i=0}^{n-1} \gamma_i^2 \left(z^{(i)} \right) \left(z^{(i)} \right)^2 - \varepsilon V_2 - \frac{a}{2} q \left[U (E) \right] U (E)$$

$$+ \frac{e^{\varepsilon d^*}}{1 - \bar{d}} \left[\Delta_2^2 (E) \, \|E\|^2 + \sum_{i=0}^{n-1} \lambda_i^2 \left(z^{(i)} \right) \left(z^{(i)} \right)^2 \right] + \widehat{\varphi}_1 \left(|z| \right) + \widehat{\varphi}_2 \left(|z| \right) . \tag{4.44}$$

According to (4.17), it can be verified that

$$\dot{V} \leq -\mu Q (W) \, W + 2\kappa + \frac{1}{2} \Theta^2 - \frac{1}{2} \widetilde{\Theta}^2 + \frac{1}{2} \Upsilon^T \Upsilon - \frac{\widetilde{\Upsilon}^T \Gamma^{-1} \widetilde{\Upsilon}}{2 \lambda_{\max} \left(\Gamma^{-1} \right)}$$

$$+ \Delta_1^2 (E) \, \|E\|^2 + \frac{e^{\varepsilon d^*}}{1 - \bar{d}} \Delta_2^2 (E) \, \|E\|^2 + \rho Q (W) \, W$$

$$- \varepsilon V_2 - \frac{a}{2} q \left[U (E) \right] U (E) + \widehat{\varphi}_1 \left(|z| \right) + \widehat{\varphi}_2 \left(|z| \right) . \tag{4.45}$$

Considering the inequality presented in Remark 4.3, one get

$$\dot{V} \leq -\mu Q (W) \, W + \rho Q (W) \, W + 2\kappa - \varepsilon V_2 + \frac{1}{2} \Theta^2 - \frac{1}{2} \widetilde{\Theta}^2$$

$$+ \frac{1}{2} \Upsilon^T \Upsilon - \frac{\widetilde{\Upsilon}^T \Gamma^{-1} \widetilde{\Upsilon}}{2 \lambda_{\max} \left(\Gamma^{-1} \right)}$$

$$+ \frac{a}{4} q \left[U (E) \right] U (E) - \frac{a}{2} q \left[U (E) \right] U (E) + \widehat{\varphi}_1 \left(|z| \right) + \widehat{\varphi}_2 \left(|z| \right) . \tag{4.46}$$

Then, we can choose a desired function $Q(W)$ satisfied

$$\widehat{\varphi}_1(|z|) + \widehat{\varphi}_2(|z|) \le \delta Q(W) W, \tag{4.47}$$

besides, let δ satisfy $\delta \le \mu - \rho$.

Combining (4.46) and (4.47), it is easy to get

$$\dot{V} \le -\mu Q(W) W + \rho Q(W) W + \delta Q(W) W + 2\kappa + \frac{1}{2}\Theta^2$$

$$-\frac{1}{2}\widetilde{\Theta}^2 + \frac{1}{2}\Upsilon^T \Upsilon - \frac{\widetilde{\Upsilon}^T \Gamma^{-1} \widetilde{\Upsilon}}{2\lambda_{\max}(\Gamma^{-1})} - \frac{a}{4}q[U(E)]U(E) - \varepsilon V_2. \tag{4.48}$$

Using (4.48), it is clear that the adaptive control law (4.15) and (4.16) can make $\dot{V}$ achieve

$$\dot{V} \le -\zeta Q(W) W - \varepsilon V_2 + 2\kappa + \frac{1}{2}\Theta^2 - \frac{1}{2}\widetilde{\Theta}^2$$

$$+\frac{1}{2}\Upsilon^T \Upsilon - \frac{\widetilde{\Upsilon}^T \Gamma^{-1} \widetilde{\Upsilon}}{2\lambda_{\max}(\Gamma^{-1})} - \frac{a}{4}q[U(E)]U(E), \tag{4.49}$$

where $\zeta = \mu - \rho - \delta \ge 0$ is a constant.

It is well known

$$\int_0^W Q(s)ds \le Q(W) W, \tag{4.50}$$

and

$$\int_0^{U(E)} q(s)ds \le q(U(E))U(E). \tag{4.51}$$

Substituting (4.50) and (4.51) into (4.49), is easy to get

$$\dot{V} \le -\frac{a}{4}\int_0^{U(E)} q(s)ds - \frac{1}{2}\widetilde{\Theta}^2 - \frac{\widetilde{\Upsilon}^T \Gamma^{-1} \widetilde{\Upsilon}}{2\lambda_{\max}(\Gamma^{-1})} - \zeta \int_0^W Q(s)ds$$

$$-\varepsilon V_2 + 2\kappa + \frac{1}{2}\Theta^2 + \frac{1}{2}\Upsilon^T \Upsilon$$

$$\le -\sigma V + \chi, \tag{4.52}$$

where $\sigma = \min\{\frac{a}{4}, \zeta, \varepsilon, \frac{1}{2}, \frac{1}{2\lambda_{\max}(\Gamma^{-1})}\}$ and $\chi = 2\kappa + \frac{1}{2}\Theta^2 + \frac{1}{2}\Upsilon^T \Upsilon$.

Multiplying both sides of (4.52) by $e^{\sigma t}$, we can obtain

$$\frac{d}{dt}\left(Ve^{\sigma t}\right) \leq \chi e^{\sigma t}. \tag{4.53}$$

Integrating both sides of (4.53) in $[0, t]$, we get

$$V \leq \left(V(0) - \frac{\chi}{\sigma}\right)e^{-\sigma t} + \frac{\chi}{\sigma}. \tag{4.54}$$

From (4.52) and (4.54), it can be concluded that the signals $E, z, \dot{z}, \ldots, z^{(n-1)}$, $\tilde{\Theta}$, and $\tilde{\Upsilon}$ remain bounded. Consequently, the adaptive controller drives the time-delay closed-loop system states to asymptotically approach a compact set, guaranteeing uniform ultimate boundedness of all closed-loop signals.

Remark 4.3 Since that $\gamma_i(\cdot)$ are known positive continuous strictly increasing functions, there always exists a positive nondecreasing continuous function $Q(\cdot)$ such that (4.17) and (4.47) holds. In addition, due to

$$\sum_{i=0}^{n-1}\left[\gamma_i^2\left(z^{(i)}\right) + \frac{e^{\varepsilon d^*}}{1 - \bar{d}}\lambda_i^2\left(z^{(i)}\right)\right]\left(z^{(i)}\right)^2$$

$$\leq \frac{W}{\lambda_{\min}(P)}\sum_{i=0}^{n-1}\left[\gamma_i^2\left(z^{(i)}\right) + \frac{e^{\varepsilon d^*}}{1 - \bar{d}}\lambda_i^2\left(z^{(i)}\right)\right], \tag{4.55}$$

we can choose

$$Q(W) \geq \frac{1}{\rho\lambda_{\min}(P)}\sum_{i=0}^{n-1}\left[\gamma_i^2\left(z^{(i)}\right) + \frac{e^{\varepsilon d^*}}{1 - \bar{d}}\lambda_i^2\left(z^{(i)}\right)\right]. \tag{4.56}$$

Remark 4.4 Compared with methodologies in Zhang et al. (2017b, 2018a), our control framework obviates backstepping requirements, eliminating repetitive virtual controller cycles while simplifying both control synthesis and closed-loop stability assessment. By employing Lyapunov-Krasovskii functionals, we construct an adaptive controller that mitigates time-delay uncertainties in HOFA systems and ensures uniform state boundedness. Diverging from Hua et al. (2008b), the proposed architecture integrates an adaptive dead-zone inverse compensator to counteract unknown input nonlinearities. This foundation enables formulation of compensation error dynamics for adaptive control design, augmented by robust control mechanisms to attenuate residual dead-zone compensation errors.

4.5 Numerical Simulations Demonstrating Controller Effectiveness

This section presents numerical simulations of a time-delay nonlinear HOFA system with unmodeled dynamics and dead-zone input to demonstrate the efficacy of the developed adaptive control framework.

We investigate a HOFA system incorporating time-delay nonlinearities, unmodeled dynamics, and an unidentified dead-zone input, described as follows:

$$\dot{e} = -e + \theta_0 z\,(t - d(t)) + z^2, \tag{4.57}$$

$$\ddot{z} = h\,(\cdot) + u\,(t), \tag{4.58}$$

$$u\,(t) = \begin{cases} 5\,(v\,(t) - 1.2), & v\,(t) \geq 1.2 \\ 0, & -2.5 < v\,(t) < 1.2 \\ 3\,(v\,(t) + 2.5), & v\,(t) \leq -2.5 \end{cases}, \tag{4.59}$$

where $\theta_0 = 1, \theta_1 = 2, \theta_2 = 3, e \in R$ is unmodeled dynamics, and z is the state variable.

By Assumption 4.1, we have

$$\begin{aligned} h\,(\cdot) &\leq 2\,|e|\,|e| + |e\,(t - d(t))| + \theta_1\,|z|\,|z| + \theta_2\,|z\,(t - 2d(t))| \\ &\quad + (\theta_2 + 1)\,|\dot{z}\,(t - d(t))| \\ &\leq R\,(\theta)\,[2\,|e|\,|e| + |e\,(t - d(t))| + |z|\,|z| + |z\,(t - 2d(t))| \\ &\quad + |\dot{z}\,(t - d(t))|], \end{aligned} \tag{4.60}$$

where $h\,(\cdot) = \theta_1 z^2 - \theta_2 z\,(t - 2d(t)) + (\theta_2 + 1)\,\dot{z}\,(t - d(t)) + 2e^2 + e\,(t - d(t))$, $d(t) = 0.5(1 + 0.2\sin(t)), R\,(\theta) = 3 + \theta_1^2 + 2\theta_2^2 = 25$.

Then, we can get that $\Delta_1\,(e) = 2\,|e|, \Delta_2\,(e_d) = 1, \gamma_{01}\,(z) = |z|, \lambda_{02}\,(z_d) = 1, \lambda_{11}\,(\dot{z}_d) = 1$.

According to the design process of the controller, the coefficient vector $A_{0\sim 1}$ is chosen as

$$A = \begin{bmatrix} 50 & 15 \end{bmatrix}, \tag{4.61}$$

$$\Phi\,(A) = \begin{bmatrix} 0 & 1 \\ -50 & -15 \end{bmatrix}. \tag{4.62}$$

Consider the Lyapunov inequation (4.16), we choose $\mu = 4$ and obtain

$$P = \begin{bmatrix} 10.2008 & 0.4280 \\ 0.4280 & 0.1098 \end{bmatrix}. \tag{4.63}$$

Fig. 4.2 Trajectory of the state $z(t)$ and $\dot{z}(t)$

Fig. 4.3 Trajectory of the state $e(t)$

Fig. 4.4 The responses of $\widehat{\Upsilon}$

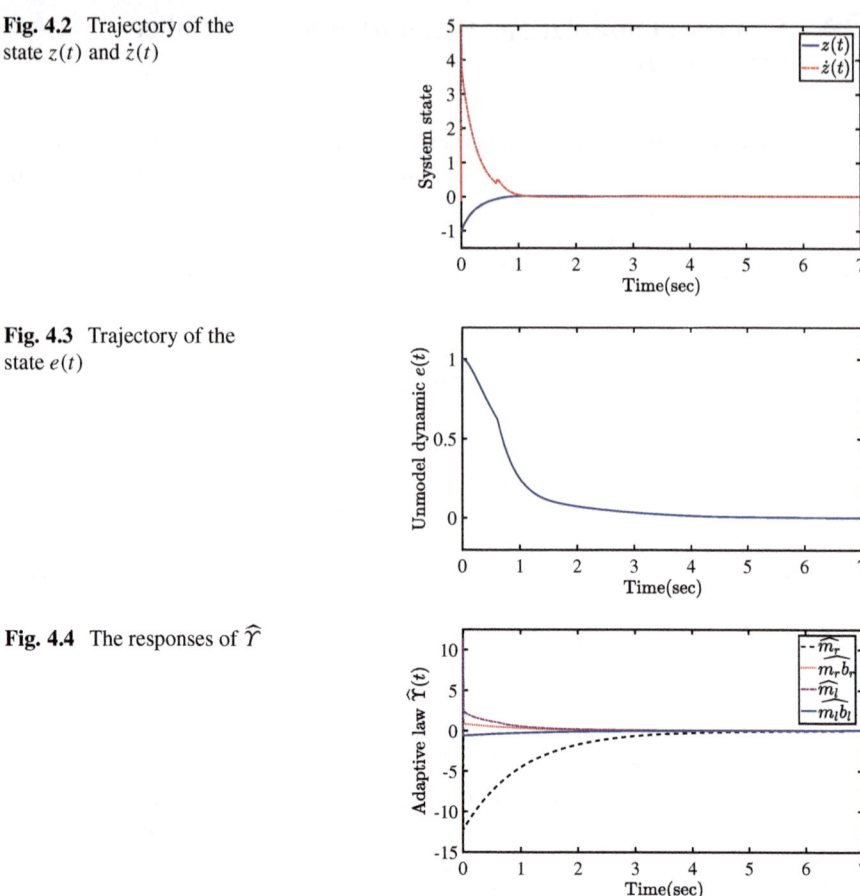

Further, we choose

$$Q(W) = 10 \left(z^{(0 \sim 1)} \right)^T P \left(z^{(0 \sim 1)} \right) + 2. \tag{4.64}$$

In addition, we choose $\Gamma \in \mathbb{R}^{4 \times 4}$ as a unit matrix, $\rho = 10$ and $\kappa = 1$.

The simulation results are shown in Figs. 4.2, 4.3, 4.4, and 4.5. Under the initial conditions $e(0) = 1, z(0) = -1, \dot{z}(0) = 1, \Theta(0) = 0$ and $\Upsilon(0) = [0.5, 0.5, -0.5, 0.5]^T$, we can see that the system state $z(t)$ and $\dot{z}(t)$ converge to origin asymptotically in Fig. 4.2. From Fig. 4.3, it can be seen that the unmodeled dynamic $e(t)$ eventually converge to zero. In addition, Fig. 4.4 shows the agility of the inverse dead-zone adaptive law $\widehat{\Upsilon}(t)$ approximation dead-zone characteristic parameters $m_r, m_r b_r, m_l$ and $m_l b_l$. Finally, one can clearly see that the adaptive parameter $\widehat{\Theta}(t)$ is bounded in Fig. 4.5. The system achieves bounded stability and verifies the feasibility of the designed adaptive controller.

Fig. 4.5 The response of
$\widehat{\Theta}(t)$

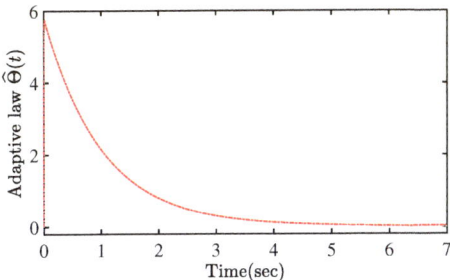

4.6 Notes and References

This chapter proposes an adaptive control framework for time-delay nonlinear HOFA systems with unmodeled dynamics and asymmetric dead-zone inputs. By integrating supply rate adjustment with the HOFA approach, unmodeled dynamics are systematically addressed, extending HOFA's applicability. An adaptive dead-zone inverse compensator is designed to comopensate for unknown asymmetric input nonlinearity, while robust control suppresses residual errors. Unlike backstepping-based methods, it is directly proven the stability via Lyapunov-Krasovskii analysis, ensuring semi-global boundedness of closed-loop states without recursive design. Simulations validate robustness against delays and nonlinearities, offering a concise solution for practical systems with actuator dead-zone.

Part II
Adaptive Time-Constrained Control for High-Order Fully Actuated Nonlinear Systems

Chapter 5
Time-Constrained Control for Strong Interconnected High-Order Fully Actuated Nonlinear Systems

This chapter focuses on the design of time-constrained controllers for HOFA nonlinear systems. By integrating the framework of practical prescribed time control, the control objectives are not only achieved within a predefined time but also ensured robust performance despite system complexities. The chapter systematically explores the theoretical foundations, controller design strategies, and stability analysis techniques required to address the challenges posed by time constraints in nonlinear systems.

5.1 Overview

In modern engineering practice, interconnected systems—such as computer networks and multi-body spacecraft architectures—represent a critical class of complex systems. Recent years have witnessed growing academic interest in control strategies for interconnected systems (Wen 1994; Jiang 2000; Wang & Lin 2015; Zhang 2022; Jin et al. 2022; Ma et al. 2021; Su et al. 2020), where decentralized control emerges as a pivotal methodology by leveraging subsystem-specific state variables in Wen (1994), Jiang (2000), and Wang and Lin (2015). While notable progress has been achieved in stabilization and tracking control for weakly coupled configurations (Zhang 2022; Jin et al. 2022; Ma et al. 2021; Su et al. 2020), current solutions predominantly address scenarios where interconnections involve only local subsystem states or neighboring outputs. Recent advancements focus on strong interconnected systems characterized by global state dependencies in both coupling terms and bounding functions. Seminal works by Zhang and Lin (2014) and Niu et al. (2022) developed decentralized controllers incorporating cross-subsystem prior knowledge, whereas Sun et al. (2020a), Li and Yang (2017), and Song et al. (2021c) achieved strict decentralization via recursive backstepping methodology and graph-theoretic frameworks. Notably, non-recursive design by

Ma and Xu (2021) was introduced asymptotic tracking for strong interconnected systems with actuator failures. This chapter extends these efforts by proposing a HOFA framework to streamline control synthesis for strong interconnected systems.

Constrained control research, particularly prescribed performance control (PPC), has gained momentum in addressing output limitations (Wang et al. 2022a; Li & Tong 2018; Wu et al. 2022; Chen et al. 2021; Liu et al. 2018; Chen 2015; Zhang et al. 2021a). Conventional PPC ensures predefined convergence rates, overshoot suppression, and steady-state error bounds (Chen 2015; Zhang et al. 2021a), with recent extensions including finite-time PPC for accelerated convergence (Liu et al. 2019) and its adaptation to unknown nonlinear systems (Zhou et al. 2021; Yuan et al. 2022). However, such methods inherently achieve semi-global stability due to initial state constraints. The practical prescribed-time control framework prooposed by Cao et al. (2022b) overcomes this limitation by decoupling convergence timing, precision specifications, and initial conditions, enabling globally valid solutions.

This chapter innovatively integrates the HOFA framework with practical prescribed-time control to address output-constrained stabilization in strong interconnected nonlinear systems. The proposed methodology inherits the structural simplicity of HOFA systems while incorporating the global performance guarantees of prescribed-time control, offering a novel paradigm for complex interconnected system regulation.

5.2 Problem Formulation and Preliminaries

5.2.1 Graph Theory

A directed graph $G = (\mathcal{V}, E)$ contains a set $\mathcal{V} = \{1, 2, \ldots, N\}$ of vertices and a set E of arcs (i, j) leading from initial vertex i to terminal vertex j. A subgraph $\mathcal{H}$ of G is said to be spanning if $\mathcal{H}$ and G have the same vertex set. A directed graph G is weighted if each arc (j, i) is assigned a non-negative weight l_{ij}. The weight $\omega(\mathcal{H})$ of a subgraph $\mathcal{H}$ is the product of the weights on all its arcs. A spanning tree of a directed graph is a subgraph on which all other nodes can be reached from a node through any path.

A directed path $\mathcal{P}$ in G is a subgraph with distinct vertices $i_1, i_2, \ldots, i_m$ such that its set of arcs is $\{(i_k, i_{k+1}) : k = 1, \ldots, m - 1\}$. If $i_m = i_1$, we call $\mathcal{P}$ a directed cycle. A connected subgraph Γ is a tree if it contains no cycles, directed or undirected. A tree Γ is rooted at vertex i, called the root, if i is not a terminal vertex of any arcs, and each of the remaining vertices is a terminal vertex of exactly one arc. A subgraph Q is unicyclic if it is a disjoint union of rooted trees whose roots form a directed cycle.

For the weighted digraph G with N vertices, define the weight matrix $\Lambda = (l_{ij})^{N \times N}$, where l_{ij} is associated with arc (j, i) of the digraph. (G, Λ) is denoted as a weighted digraph to model the topology of all links in the network. A directed graph G is strongly connected if there exists a directed path from one to the other for any pair of different nodes. A weighted digraph $(G, \varLambda)$ is strongly connected if and only if the weight matrix $\varLambda$ is irreducible. The Laplacian matrix of $(G, \varLambda)$ is defined as

$$
L = \begin{bmatrix}
\sum_{k \neq 1} l_{1k} & -l_{12} & \cdots & -l_{1N} \\
-l_{21} & \sum_{k \neq 2} l_{2k} & \cdots & -l_{2N} \\
\vdots & \vdots & \ddots & \vdots \\
-l_{N1} & -l_{N2} & \cdots & \sum_{k \neq N} l_{Nk}
\end{bmatrix}.
\tag{5.1}
$$

5.2.2   System Formulation for Strong Interconnected High-Order Fully Actuated Nonlinear Systems

This chapter investigates interconnected nonlinear systems structured over a directed graph G, where each subsystem resides at a node and interacts with neighboring subsystems through the graph's arcs. For the i-th subsystem ($i \in \mathcal{V}$), the dynamics are characterized as:

$$
\begin{cases}
\dot{x}_{ij}(t) = x_{ij+1}, \\
\dot{x}_{in}(t) = g_i(\bar{x}_{in}, t)u_i + f_i(\bar{x}_{in}, t) + h_{in}(\bar{x}_{1n}, \bar{x}_{2n}, \ldots, \bar{x}_{Nn}), \\
y_i(t) = x_{i1},
\end{cases}
\tag{5.2}
$$

where $1 \leq i \leq N$, $1 \leq j \leq n - 1$; $x_{ij}(t) \in \mathbb{R}$, $u_i(t) \in \mathbb{R}$ and $y_i(t) \in \mathbb{R}$ are the state variables, control input and output of the ith subsystem. $\bar{x}_{ij} = [x_{i1}(t), x_{i2}(t), \cdots, x_{ij}(t)]^T$. $g_i(\bar{x}_{in}, t)$ and $f_i(\bar{x}_{in}, t)$ are known smooth nonlinear functions with $g_i(\bar{x}_{in}, t) \neq 0$ and $f_i(0, t) = 0$. h_{in} is an unknown strong interconnected nonlinear function with $h_{ij}(0) = 0$.

The control objective is to design a controller u_i such that:

(1) the output of each subsystem can converge to a predefined set in a prescribed time, which is independent of the initial condition.
(2) all signals of the closed-loop system are globally bounded.

To further facilitate the controller design, the following assumptions and lemmas are given.

Assumption 5.1 (Li and Yang (2017)) The interconnection term satisfies

$$
|h_{in}(\bar{x}_{1n}, \bar{x}_{2n}, \ldots, \bar{x}_{Nn})| \leq \sum_{j=1}^{N} \phi_{jn}(\bar{x}_{jn}),
\tag{5.3}
$$

where $\phi_{jn}(\bar{x}_{jn})$ is a known continuous nonlinear function.

Assumption 5.2 (Li and Yang (2017)) The digraph is strongly connected.

Lemma 5.1 (Li and Shuai (2010)) *Suppose that $N \geq 2$. Then*

$$\beta_i = \sum_{\Gamma \in T_i} \omega(\Gamma), \ i \in \mathcal{V},$$

where β_i is the cofactor of the ith diagonal element of L. If the directed graph G is strongly connected, $\beta_i > 0$. T_i is the set of all spanning trees Γ of (G, Λ) that are rooted at vertex i, and $\omega(\Gamma)$ is the weight of Γ.

Lemma 5.2 (Li and Shuai (2010)) *Suppose that $N \geq 2$. Let β_i be given in Lemma 5.1. Then*

$$\sum_{i=1}^{n} \sum_{j=1}^{n} \beta_i l_{ij} H_{ij}(x_i, x_j) = \sum_{Q \in \mathbb{Q}} \omega(Q) \sum_{(s,r) \in E(C_Q)} H_{rs}(x_r, x_s),$$

where $H_{ij}(x_i, x_j)$ $(i, j \in \mathcal{V})$ are arbitrary functions. $\omega(Q)$ is the weight of Q, and C_Q is denoted as the directed cycle of Q. $\mathbb{Q}$ is the set of all spanning unicyclic graphs of (G, Λ).

According to Lemmas 2.5 and 2.6, it can be deduced that, for any $\mu > 0$, there is a positive definite matrix $P(A^{0 \sim n-1})$ such that $\Phi^T(A^{0 \sim n-1}) P(A^{0 \sim n-1}) + P(A^{0 \sim n-1}) \Phi(A^{0 \sim n-1}) \leq -\mu P(A^{0 \sim n-1})$. Then we define

$$P_L(A^{0 \sim n-1}) = P(A^{0 \sim n-1}) \begin{bmatrix} 0_{n-1} \\ I \end{bmatrix},$$

which will be used in the latter analysis.

Proposition 5.1 (Duan (2021e)) *For any selected $F \in \mathbb{R}^{n \times n}$, all matrices $A^{0 \sim n-1}$ and $V \in \mathbb{R}^{n \times n}$ satisfying $\det V \neq 0$ and $\Phi(A^{0 \sim n-1}) = VFV^{-1}$ are derived from the equation $A^{0 \sim n-1} = -ZF^n V^{-1}(Z, F)$ with*

$$V(Z, F) = \begin{bmatrix} Z \\ ZF \\ \dots \\ ZF^{n-1} \end{bmatrix},$$

where $Z \in \mathbb{R}^{1 \times n}$ is a designed parameter matrix satisfying $\det V(Z, F) \neq 0$.

Remark 5.1 Assumption 5.1 establishes a bounding constraint for strong interconnection terms through nonlinear functions, reflecting a commonly employed generic assumption in networked system studies. Though existing approaches (Niu et al. 2022; Sun et al. 2020a) mitigate this constraint by permitting unknown strong interconnections governed by unspecified functions, such configurations achieve only semi-global stabilization—a fundamental distinction from our globally convergent framework. Assumption 5.2 derives its justification from the persistent coupling effect between neighboring subsystems. Proposition 5.1 operationalizes the closed-loop system's linear parametrization process, extending the parametric construction principles established in Duan (2020a) and Duan (2021e).

Remark 5.2 The Laplacian matrix L is uniquely determined by Assumption 5.1, where entries l_{ij} binary-code the existence of $\phi_{jn}(\bar{x}_{jn})$. For subsystem i, if h_{in} admits decomposition into summation terms $\phi_{1n}(\bar{x}_{1n}) + \cdots + \phi_{Nn}(\bar{x}_{Nn})$ with identifiable $\phi_{jn}(\bar{x}_{jn})$, then $l_{ij} = 1$ in L; otherwise $l_{ij} = 0$. This binarized encoding scheme enables systematic construction of L.

5.2.3   Practical Prescribed-Time Control

To ensure convergence of system outputs to a predefined set within specified time constraints, the practical prescribed-time control strategy from Cao et al. (2022b) is implemented. This control objective is formally defined as

$$- \rho_i(t) < y_i(t) < \rho_i(t), \tag{5.4}$$

where $\rho_i(t)$ is the performance function and designed as

$$\rho_i(t) = \begin{cases} \left(\frac{1}{t} - \frac{1}{T_f}\right)^{2p} + \rho_{T_f}, & t \in (0, T_f], \\ \rho_{T_f}, & t \in (T_f, \infty), \end{cases} \tag{5.5}$$

with ρ_{T_f} and T_f being the positive designed parameters, p being a positive integer that satisfies $2p > n + 1$. Then, a compound function $h_i(\gamma_i(t))$ is constructed as

$$h_i(\gamma_i(t)) = \begin{cases} 1 - \left(\frac{\gamma_i(t)}{a} - 1\right)^{2p}, & 0 < \gamma_i(t) \le a, \\ 1, & \gamma_i(t) > a, \end{cases} \tag{5.6}$$

where $a > c\rho_{T_f}^2$, $\gamma_i(t) = c(\rho_i(t) - y_i(t))(\rho_i(t) + y_i(t))$ with $c > 0$. To proceed, the following transformation is given

$$s_{i1}(t) = \frac{y_i(t)}{h_i(\gamma_i(t))}, \tag{5.7}$$

from which we can get some properties: (i) $s_{i1}(t) = 0$ only when $y_i(t) = 0$; (ii) $s_{i1}(t) \to \infty$ as $y_i(t) \to \rho_i(t)$ or $y_i(t) \to -\rho_i(t)$; (iii) when $\gamma_i(t) > a$, the system is operating in safe region, and $s_{i1}(t) = y_i(t)$ due to $h_i(\gamma_i(t)) = 1$.
   Then it follows that

$$\dot{s}_{i1} = r_{i1}\dot{y}_i + v_{i1} = r_{i1}x_{i2} + v_{i1}, \tag{5.8}$$

where

$$r_{i1} = \begin{cases} \frac{1}{h_i} - \frac{4cp}{ah_i^2} \left(\frac{\gamma_i}{a} - 1 \right)^{2p-1} y_i^2, & 0 < \gamma_i(t) \le a, \\ 1, & \gamma_i(t) > a, \end{cases} \tag{5.9}$$

and

$$v_{i1} = \begin{cases} \frac{4cp}{ah_i^2} \left(\frac{\gamma_i}{a} - 1 \right)^{2p-1} \rho_i \dot{\rho}_i y_i, & 0 < \gamma_i(t) \le a, \\ 0, & \gamma_i(t) > a, \end{cases} \tag{5.10}$$

from which we can see that r_{i1} is strictly positive and computable.

Define

$$s_{ij} = x_{ij}, \, j = 2, \ldots, n, \tag{5.11}$$

then the system (5.2) can be transformed into the following form

$$\begin{cases} \dot{s}_{i1} = g_i^* s_{i2} + f_i^*, \\ \dot{s}_{ij} = s_{ij+1}, \quad j = 2, \ldots, n-1, \\ \dot{s}_{in} = g_i(\bar{x}_{in}, t) u_i + f_i(\bar{x}_{in}, t) + h_{in}(\bar{x}_{1n}, \bar{x}_{2n}, \ldots, \bar{x}_{Nn}), \end{cases} \tag{5.12}$$

where

$$g_i^*(s_{i1}, t) = r_{i1}, \, f_i^*(s_{i1}, t) = v_{i1}. \tag{5.13}$$

Remark 5.3 The definition of $\rho_i(t)$ yields $\rho_i(0) = \infty$, inherently guaranteeing satisfaction of the initial constraint $|y_i(0)| < \rho_i(0)$ for any bounded $y_i(0)$. Through transformation (5.7), the control singularity caused by infinite $\gamma_i(0)$ is effectively resolved. This methodology successfully eliminates the restrictive initial condition requirements imposed in Chen (2015), Zhang et al. (2021a), Liang et al. (2020), Xu (2021), Sun et al. (2022), and Liu et al. (2019).

5.3 Time-Constrained Decentralized Controller Design Based on Practical Prescribed-Time Control

This section develops a decentralized control scheme through the HOFA system framework. The design initiates with the proposed coordinate transformation:

$$z_i = s_{i1},$$

$$s_{i2} = \left(g_i^* \right)^{-1} (\dot{z}_i - f_i^*), \tag{5.14}$$

then we can obtain

$$\ddot{z}_i = \dot{g}_i^* s_{i2} + g_i^* \dot{s}_{i2} + \dot{f}_i^* = g_i^* s_{i3} + F_{i2}, \tag{5.15}$$

where

$$F_{i2}(z_i, \dot{z}_i, t) = \dot{g}_i^* s_{i2} + \dot{f}_i^*. \tag{5.16}$$

From Eq. (5.15), it is easy to obtain

$$s_{i3} = \left(g_i^*\right)^{-1}(\ddot{z}_i - F_{i2}), \tag{5.17}$$

then, by taking the derivative of $\ddot{z}_i$ w.r.t (with respect to) time, we get

$$z_i^{(3)} = \dot{g}_i^* s_{i3} + g_i^* \dot{s}_{i3} + \dot{F}_{i2} = g_i^* s_{i4} + F_{i3}, \tag{5.18}$$

where

$$F_{i3}(z_i, \dot{z}_i, z_i^{(2)}, t) = \dot{F}_{i2} + \dot{g}_i^* s_{i3}. \tag{5.19}$$

Similar to this design procedure, we can obtain

$$s_{ij} = \left(g_i^*\right)^{-1}(z_i^{(j-1)} - F_{ij-1}), \quad j = 4, \ldots, n-1, \tag{5.20}$$

then the following high-order system form can be obtained

$$z_i^{(j)} = \dot{g}_i^* s_{ij} + g_i^* \dot{s}_{ij} + \dot{F}_{ij-1} = g_i^* s_{ij+1} + F_{ij}, \tag{5.21}$$

where

$$F_{ij}(z_i, \dot{z}_i, \ldots, z_i^{(j-1)}, t) = \dot{F}_{ij-1} + \dot{g}_i^* s_{ij}. \tag{5.22}$$

When $j = n$, we have

$$
\begin{aligned}
z_i^{(n)} &= \dot{g}_i^* s_{in} + g_i^* \dot{s}_{in} + \dot{F}_{in-1} \\
&= g_i^* g_i(\bar{x}_{in}, t)u_i + g_i^* h_{in}(\bar{x}_{1n}, \bar{x}_{2n}, \ldots, \bar{x}_{Nn}) + F_i \\
&= G_i u_i + H_i + F_i,
\end{aligned} \tag{5.23}
$$

where

$$
\begin{aligned}
G_i &= g_i^* g_i(\bar{x}_{in}, t), \\
H_i &= g_i^* h_{in}(\bar{x}_{1n}, \bar{x}_{2n}, \ldots, \bar{x}_{Nn}),
\end{aligned} \tag{5.24}
$$

$$F_i(z_i, \dot{z}_i, \dots, z_i^{(n-1)}, t) = \dot{F}_{in-1} + \dot{g}_i^* s_{in} + g_i^* f_i(\bar{x}_{in}, t). \qquad (5.25)$$

According to the work Duan (2021e), the controller for the above HOFA system (5.23) is designed as

$$u_i = -G_i^{-1}(u_{i0} + u_{i1}), \qquad (5.26)$$

where

$$
\begin{aligned}
u_{i0} &= A^{0\sim n-1} z_i^{(0\sim n-1)} + F_i, \\
u_{i1} &= \tfrac{r_{i1}^2}{4\varepsilon_i} \sum_{j=1}^{N} \phi_{in}^2(\bar{x}_{in}) P_L^T z_i^{(0\sim n-1)},
\end{aligned} \qquad (5.27)
$$

where ε_i is the positive design parameter.

Substituting the control law (5.26) into the system (5.23) gives the following closed-loop system

$$z_i^{(n)} + A^{0\sim n-1} z_i^{(0\sim n-1)} = H_i - u_{i1} \qquad (5.28)$$

The closed-loop system (5.28) can be written in the following state-space form

$$\dot{z}_i^{(0\sim n-1)} = \Phi(A^{0\sim n-1}) z_i^{(0\sim n-1)} + \begin{bmatrix} 0_{n-1} \\ H_i - u_{i1} \end{bmatrix}. \qquad (5.29)$$

Then the following Lyapunov function can be chosen for the above system

$$V_i = \frac{1}{2} \left(z_i^{(0\sim n-1)} \right)^T P(A^{0\sim n-1}) z_i^{(0\sim n-1)}. \qquad (5.30)$$

According to the properties of Lemmas 2.5 and 2.6, we can get

$$
\begin{aligned}
\dot{V}_i &= \frac{1}{2} \left(\dot{z}_i^{(0\sim n-1)} \right)^T P(A^{0\sim n-1}) z_i^{(0\sim n-1)} + \frac{1}{2} \left(z_i^{(0\sim n-1)} \right)^T P(A^{0\sim n-1}) \dot{z}_i^{(0\sim n-1)} \\
&= \frac{1}{2} \left(z_i^{(0\sim n-1)} \right)^T (\Phi^T(A^{0\sim n-1}) P(A^{0\sim n-1}) + P(A^{0\sim n-1}) \Phi(A^{0\sim n-1})) z_i^{(0\sim n-1)} \\
&\quad + \left(z_i^{(0\sim n-1)} \right)^T P(A^{0\sim n-1}) \begin{bmatrix} 0_{n-1} \\ H_i - u_{i1} \end{bmatrix} \\
&\leq -\mu V_i + \left(z_i^{(0\sim n-1)} \right)^T P_L(A^{0\sim n-1})(H_i - u_{i1}). \qquad (5.31)
\end{aligned}
$$

From Eqs. (5.24) and (5.27), and Assumption 5.1, we have

$$\left(z_i^{(0\sim n-1)}\right)^T P_L(H_i - u_{i1})$$

$$\leq \left\|\left(z_i^{(0\sim n-1)}\right)^T P_L\right\| \sum_{j=1}^N r_{i1}\phi_{jn}(\bar{x}_{jn}) - \left(z_i^{(0\sim n-1)}\right)^T P_L u_{i1}$$

$$\leq \left\|\left(z_i^{(0\sim n-1)}\right)^T P_L\right\| \sum_{j=1}^N r_{i1}\phi_{jn}(\bar{x}_{jn}) - \left(z_i^{(0\sim n-1)}\right)^T$$

$$P_L \frac{r_{i1}^2}{4\varepsilon_i} \sum_{j=1}^N \phi_{in}^2(\bar{x}_{in}) P_L^T z_i^{(0\sim n-1)}$$

$$\leq \left\|\left(z_i^{(0\sim n-1)}\right)^T P_L\right\| \sum_{j=1}^N r_{i1}\phi_{jn}(\bar{x}_{jn}) - \frac{r_{i1}^2}{4\varepsilon_i}\left\|\left(z_i^{(0\sim n-1)}\right)^T P_L\right\|^2 \sum_{j=1}^N \phi_{jn}^2(\bar{x}_{jn})$$

$$+ \frac{r_{i1}^2}{4\varepsilon_i}\left\|\left(z_i^{(0\sim n-1)}\right)^T P_L\right\|^2 \left(\sum_{j=1}^N (\phi_{jn}^2(\bar{x}_{jn}) - \phi_{in}^2(\bar{x}_{in}))\right)$$

$$\leq \frac{r_{i1}^2}{4\varepsilon_i}\left\|\left(z_i^{(0\sim n-1)}\right)^T P_L\right\|^2 \left(\sum_{j=1}^N (\phi_{jn}^2(\bar{x}_{jn}) - \phi_{in}^2(\bar{x}_{in}))\right) + n\varepsilon_i. \tag{5.32}$$

Now, the whole Lyapunov function is chosen as $V = \sum_{i=1}^N \beta_i V_i$, where β_i is denoted in Lemma 5.1. From Assumption 5.2 and Lemma 5.1, we can obtain $\beta_i > 0$, then it follows that

$$\dot{V} \leq -\mu V + \sum_{i=1}^N \beta_i n\varepsilon_i + \sum_{i=1}^N \beta_i \frac{r_{i1}^2}{4\varepsilon_i}\left\|\left(z_i^{(0\sim n-1)}\right)^T P_L\right\|^2$$

$$\left(\sum_{j=1}^N (\phi_{jn}^2(\bar{x}_{jn}) - \phi_{in}^2(\bar{x}_{in}))\right). \tag{5.33}$$

According to the Lemmas 5.1 and 5.2 and the work of Li and Yang (2017), we have

$$\sum_{i=1}^N \sum_{j=1}^N \beta_i(\phi_{jn}^2(\bar{x}_{jn}) - \phi_{in}^2(\bar{x}_{in}))$$

$$= \sum_{Q\in\mathbb{Q}} \omega(Q) \sum_{(s,r)\in E(C_Q)} (\phi_{sn}^2(\bar{x}_{sn}) - \phi_{rn}^2(\bar{x}_{rn})). \tag{5.34}$$

Without losing generality, for any directed cycle C_Q, the set $E(C_Q)$ can be described as

$$E(C_Q) = \{(i_k, i_{k+1}) \mid k = 1, \ldots, m-1, m \le N, i_m = i_1\}. \tag{5.35}$$

Then, we have

$$\sum_{(s,r)\in E(C_Q)} (\phi_{sn}^2(\bar{x}_{sn}) - \phi_{rn}^2(\bar{x}_{rn})) = (\phi_{i_1,n}^2(\bar{x}_{i_1,n}) - \phi_{i_2,n}^2(\bar{x}_{i_2,n}))$$

$$+ (\phi_{i_2,n}^2(\bar{x}_{i_2,n}) - \phi_{i_3,n}^2(\bar{x}_{i_3,n})) + \ldots + (\phi_{i_m,n}^2(\bar{x}_{i_m,n}) - \phi_{i_1,n}^2(\bar{x}_{i_1,n})) = 0, \tag{5.36}$$

which implies that

$$\sum_{i=1}^{N}\sum_{j=1}^{N} \beta_i (\phi_{jn}^2(\bar{x}_{jn}) - \phi_{in}^2(\bar{x}_{in})) = 0. \tag{5.37}$$

It follows that

$$\dot{V} \le -\mu V + \sum_{i=1}^{N} \beta_i n \varepsilon_i \le -\mu V + \varepsilon, \tag{5.38}$$

with $\varepsilon = \sum_{i=1}^{N} \beta_i n \varepsilon_i$.

The main results and analysis procedures are summarized as follows.

Theorem 5.1 *Consider the interconnected nonlinear systems (5.2), and suppose that Assumptions 5.1 and 5.2 hold, then the controller shown in (5.26) ensures that all the signals of the system are globally bounded.*

Proof Multiplying (5.38) by $e^{\mu t}$, one has $d(V(t)e^{\mu t})/dt \le \varepsilon e^{\mu t}$, and integrating it over $[0, t]$ yields

$$V(t) \le \left(V(0) - \frac{\varepsilon}{\mu}\right) e^{-\mu t} + \frac{\varepsilon}{\mu}, \tag{5.39}$$

which implies $V(t)$ is bounded at the time period $[0, \infty)$. And $V \to \frac{\varepsilon}{\mu}$, as $t \to \infty$. Due to $V = \sum_{i=1}^{N} \beta_i V_i$, thus, $z_i^{(0\sim n-1)}$ eventually converges into the ellipsoid $\Theta_{(\mu,\varepsilon)}(0) = \{z_i^{(0\sim n-1)} \mid \left\|\left(z_i^{(0\sim n-1)}\right)\right\|_P^2 \le 2\frac{\varepsilon}{\beta_i \mu}\}$.

By (5.39), we can obtain that s_{i1} is bounded, which means $|y_i(t)| < \rho_i(t)$, and $|y_i(t)| < \rho_{T_f}$ after $t > T_f$. Thus, the system outputs are within the prescribed bounds and converge to the predetermined invariant region in the settling time T_f regardless of the initial conditions. Meanwhile, it is known that $z_i^{(1)}$ is bounded, and then $s_{i2}(x_{i2})$ is bounded. In this case, the boundedness of $s_{ij}(x_{ij})\ i = 1, \ldots, n,\ j = 2, \ldots, n$ can be ensured. Then, the control input u_i is also bounded. Besides, both the prescribed bound ρ_{T_f} and the settling time T_f can be predesigned. Therefore, all the signals of the closed-loop system are bounded.

Remark 5.4 In contrast to conventional approaches in Zhang and Lin (2014), Niu et al. (2022), Sun et al. (2020a), and Li and Yang (2017) that employ recursive backstepping frameworks to synthesize control laws, this work adopts a HOFA system paradigm for strong interconnected nonlinear systems. The proposed methodology involves (i) reformulating the original nonlinear dynamics into a HOFA structure, (ii) enabling direct decentralized controller synthesis without recursive virtual controller derivations, and (iii) systematic cancellation of nonlinear cross-coupling effects. As demonstrated in the analytical framework, this strategy not only enhances computational tractability but also circumvents the inherent complexity of iterative stabilization mechanisms.

Remark 5.5 In inequality (5.32), by substituting the controller and scaling of inequalities, the strong interconnected term can be reconstructed into the form of the first term on the right side of the inequality. By using the result of graph theory, we construct the Lyapunov function as $V = \sum_{i=1}^{N} \beta_i V_i$, and the first term can be eliminated according to the above analysis, which handles the strong interconnected term.

Remark 5.6 The system's operational efficiency can be enhanced through three primary parametric modifications. First, optimizing the performance function $\rho_i(t)$ and compound function $h_i(\gamma_i(t))$ by shortening the terminal time T_f accelerates the output convergence rate, while decreasing ρ_{T_f} tightens the prescribed boundary. Simultaneously, tuning parameters a and c maintains subsystem outputs within safety margins away from constraint thresholds. Second, diminishing the error coefficient ε_i contributes to performance refinement. Third, elevating the stability margin μ for the linear component strengthens system robustness. Experimental verification confirms the first approach yields the most substantial performance improvement.

5.4   Simulation Example

To evaluate the effectiveness and efficiency of the proposed control scheme, we simulate the following interconnected inverted pendulum systems according to Sun et al. (2020a)

$$
\begin{cases}
\dot{x}_{11}(t) = x_{12}(t), \\
\dot{x}_{12}(t) = \frac{1}{J_1}u_1(t) + \left(\frac{m_1 g r}{J_1} - \frac{kr^2}{4J_1}\right)\sin(x_{11}(t)) \\
\qquad + \frac{kr}{2J_1}(l-b) + \frac{kr^2}{4J_1}\sin(x_{21}(t)), \\
\dot{x}_{21}(t) = x_{22}(t), \\
\dot{x}_{22}(t) = \frac{1}{J_2}u_2(t) + \left(\frac{m_2 g r}{J_2} - \frac{kr^2}{4J_2}\right)\sin(x_{21}(t)) \\
\qquad + \frac{kr}{2J_2}(l-b) + \frac{kr^2}{4J_2}\sin(x_{11}(t)), \\
y_1(t) = x_{11}(t), \quad y_2(t) = x_{21}(t),
\end{cases}
\tag{5.40}
$$

where $x_{11}(t)$ and $x_{12}(t)$ are angular displacements of the pendulums from vertical, the parameters $J_1 = 5\,\mathrm{kg\cdot m^2}$ and $J_2 = 6.25\,\mathrm{kg\cdot m^2}$ are the inertia moments, $m_1 = 2\,\mathrm{kg}$ and $m_2 = 2.5\,\mathrm{kg}$ are the pendulum end masses, $k = 100\,\mathrm{N/m}$ is the spring constant, $l = 0.5\,\mathrm{m}$ is the spring natural length, $r = 0.5\,\mathrm{m}$ is the pendulum height, $b = 0.5\,\mathrm{m}$ is distance between the pendulum hinges, and $g = 9.81\,\mathrm{m/s^2}$ denotes the gravitational acceleration.

Based on the above strategy designed, the performance function $\rho_i(t)$ and the compound function $h_i(\gamma_i(t))$ are designed as:

$$
\rho_i(t) = \begin{cases}
\left(\frac{1}{t} - \frac{1}{T_f}\right)^{2p} + \rho_{T_f}, & t \in (0, T_f], \\
\rho_{T_f}, & t \in (T_f, \infty),
\end{cases}
\tag{5.41}
$$

and

$$
h_i(\gamma_i(t)) = \begin{cases}
1 - \left(\frac{\gamma_i(t)}{a} - 1\right)^{2p}, & 0 < \gamma_i(t) \le a, \\
1, & \gamma_i(t) > a,
\end{cases}
\tag{5.42}
$$

where $\rho_{T_f} = 0.02$, $T_f = 1.5s$, $p = 2$, $c = 50$, and $a = 5$. The controllers are designed as

$$
\begin{aligned}
u_i &= -G_i^{-1}(u_{i0} + u_{i1}), \\
u_{i0} &= A^{0\sim 1} z_i^{(0\sim 1)} + F_i, \\
u_{i1} &= \frac{r_{i1}^2}{4\varepsilon_i}\sum_{j=1}^{2}\phi_{i2}^2(\bar{x}_{i2})P_L^T z_i^{(0\sim 1)},
\end{aligned}
\tag{5.43}
$$

where $\phi_{12}(\bar{x}_{12}) = |sin(x_{11})|$, $\phi_{22}(\bar{x}_{22}) = |1.25sin(x_{21})|$. Obviously, there exist interconnections between the two adjacent subsystems, and the nonlinear functions

h_{i2} ($i = 1, 2$) satisfy the following inequalities

$$|h_{12}| \leq |1.25sin(x_{21})|, \; |h_{22}| \leq |sin(x_{11})|, \tag{5.44}$$

hence, it can be derived that the above interconnected inverted pendulum systems satisfy Assumptions 5.1 and 5.2.

Then we can take the design parameter $\varepsilon_i = 0.02$. The solution of $A^{0\sim1}$ and $P_L^T(A^{0\sim1})$ can be derived in the following procedure. Choose

$$F = \begin{bmatrix} -d_1 & -d_2 \\ d_2 & -d_1 \end{bmatrix}, \tag{5.45}$$

where d_1 and d_2 are two positive scalars. It is obvious that $-d_1 \pm d_2 j$ are the eigenvalues of the matrix F, and we set $d_1 = 3, d_2 = 1.5$. By choosing $Z = [1, 1]$ and using Proposition 5.1, we have

$$V = \begin{bmatrix} Z \\ ZF \end{bmatrix} = \begin{bmatrix} 1 & 1 \\ -1.5 & -4.5 \end{bmatrix}, \tag{5.46}$$

then we can obtain

$$A^{0\sim1} = -ZF^2V^{-1} = \begin{bmatrix} 11.25 & 6 \end{bmatrix}, \tag{5.47}$$

and it can be seen that the matrix

$$\Phi(A^{0\sim1}) = VFV^{-1} = \begin{bmatrix} 0 & 1 \\ -11.25 & -6 \end{bmatrix} \tag{}$$

has the eigenvalues $-d_1 \pm d_2 j$.

Consider the Lyapunov equation

$$(\Phi(A^{0\sim1}) + 2I)^T P(A^{0\sim1}) + P(A^{0\sim1})(\Phi(A^{0\sim1}) + 2I) = -10^{-3}I, \tag{5.48}$$

which clearly gives $\Phi^T(A^{0\sim1})P(A^{0\sim1}) + P(A^{0\sim1})\Phi(A^{0\sim1}) \leq -\mu P(A^{0\sim1})$ with $\mu = 4$. By solving the above equation, we obtain

$$P(A^{0\sim1}) = \begin{bmatrix} 0.0112 & 0.002 \\ 0.002 & 0.0006 \end{bmatrix}, \tag{5.49}$$

which means

$$P_L^T(A^{0\sim1}) = \begin{bmatrix} 0.002 & 0.0006 \end{bmatrix}. \tag{5.50}$$

Fig. 5.1 The responses of state x_{11} under the performance function

Fig. 5.2 The responses of state x_{21} under the performance function

Fig. 5.3 The responses of state x_{12}

Fig. 5.4 The responses of state x_{22}

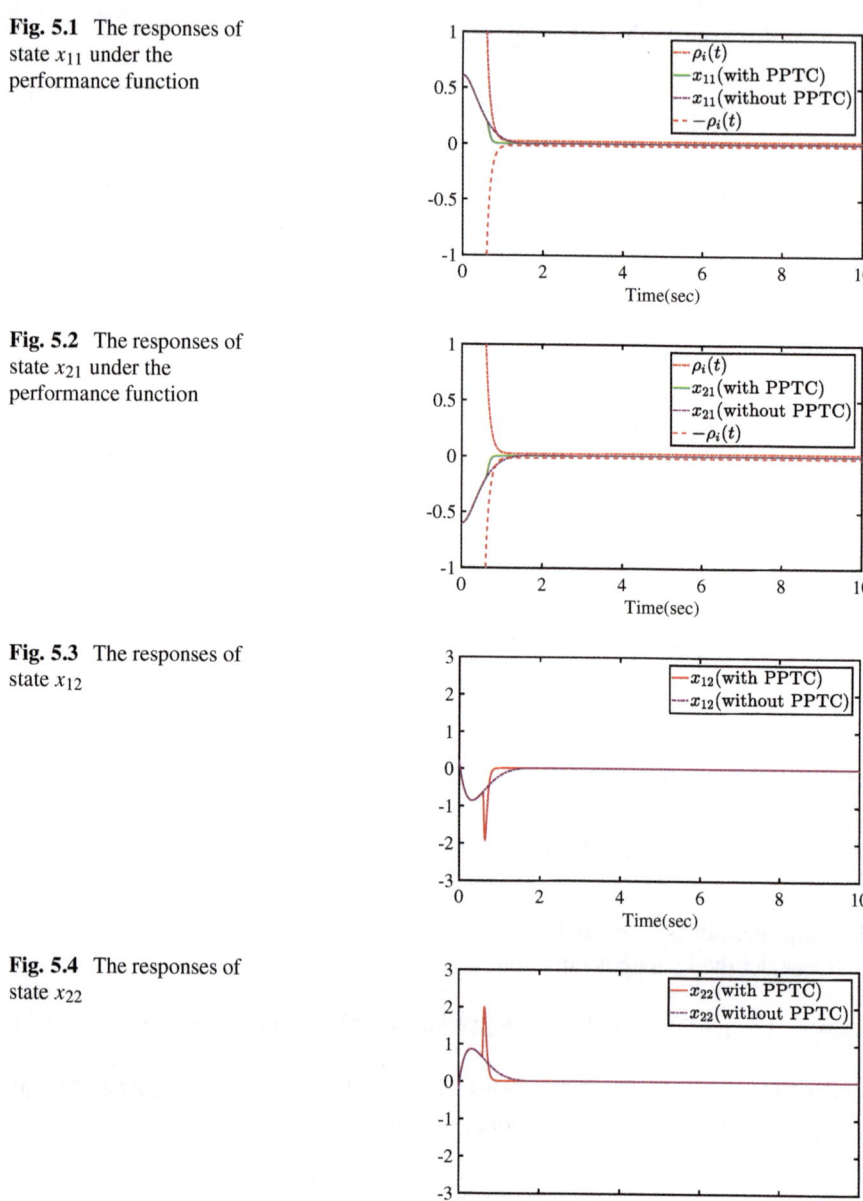

The initial conditions are given as $[x_{11}, (0)x_{12}, (0)x_{13}, (0)x_{14}(0)]^T = [0.6, 0.2, -0.6, -0.2]^T$.

Then the simulation outcomes are displayed in Figs. 5.1, 5.2, 5.3, 5.4, and 5.5. Figures 5.1 and 5.2 demonstrate that all subsystem outputs remain bounded

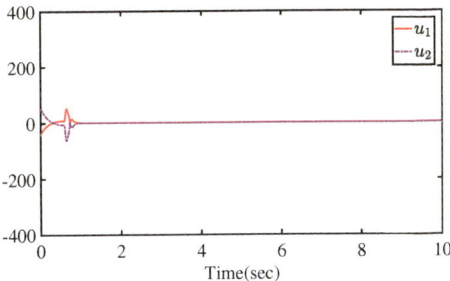

Fig. 5.5 The responses of control inputs u_1 and u_2

by the designated performance functions, confirming simultaneous satisfaction of prescribed transient boundaries and convergence to invariant regions within settling time. Notably, no constraints are imposed on initial states $x_{11}(t)$ and $x_{21}(t)$. Compared to PPTC-free strategies, the proposed method achieves accelerated convergence with enhanced precision. The state trajectories $x_{12}(t)$ and $x_{22}(t)$ in Figs. 5.3 and 5.4 exhibit asymptotic stabilization near zero. Relative to non-PPTC approaches, transient state magnitudes briefly increase but stabilize more rapidly. Control input profiles $u_1(t)$ and $u_2(t)$ for both subsystems are plotted in Fig. 5.5, revealing peak control efforts when outputs $y_i(t)$ approach constraint boundaries $\rho_i(t)$ near 0.6 s—a physically reasonable phenomenon. These collective results validate the efficacy of our control framework for interconnected inverted pendulum systems.

5.5 Notes and References

This chapter extends the HOFA system framework to address decentralized control of strong interconnected nonlinear systems with time constraints—a novel application broadening HOFA's scope. By transforming the system into HOFA form, a practical prescribed-time control strategy is proposed, ensuring outputs converge to a prescribed invariant region within a settling time without requiring prior knowledge of initial values. Unlike traditional backstepping methods, this approach eliminates repetitive virtual controller design, significantly reducing computational complexity. Leveraging graph theory, global boundedness of all closed-loop signals is rigorously proven, while adjustable performance functions enhance output flexibility compared to linear-focused designs. Simulations validate the method's efficacy, offering a streamlined solution for constrained interconnected systems and paving the way for future extensions to full-state constraints.

Chapter 6
Adaptive Time-Constrained Control for High-Order Fully Actuated Nonlinear Systems

This chapter delves into the prescribed-time stabilization of fully actuated nonlinear systems by introducing a novel adjustment function and a solution-parameter Lyapunov function. The control scheme presented offers a more straightforward and efficient alternative to conventional backstepping-based methods. Furthermore, for systems with uncertain parameters, we develop an adaptive control strategy that employs a time-varying adaptive controller, ensuring stability within a pre-determined time frame. The proposed approach is validated through a simulation example, highlighting its effectiveness.

6.1 Overview

Recent scholarly attention has increasingly focused on fully actuated system theory due to its ability to address complex control challenges with greater flexibility and efficiency. This shift marks a departure from traditional first-order system methods, which rely heavily on state-space models and struggle with high-dimensional nonlinear systems. Duan (2021a) conducted a rigorous analysis of the limitations of first-order state space method and introduced HOFA systems theory for the first time. This framework enhances the design and implementation of advanced control strategies by leveraging additional degrees of freedom, enabling more precise and reliable outcomes.

Building on Lyapunov stability theory, various robust control schemes have been developed to improve the stability and tracking performance of HOFA systems, even in the presence of nonlinear uncertainties. For example, Duan (2021c) proposed robust stabilization and tracking control schemes tailored for single fully actuated systems with inherent nonlinear uncertainties. These schemes ensure system stability and accurate trajectory tracking across diverse operating conditions, extending the practical applicability of high-order control. Additionally, to address parametric

C. Hua et al., *Adaptive Constrained Control for High-Order Fully Actuated Nonlinear Systems*, https://doi.org/10.1007/978-981-95-0962-1_6

uncertainties, Duan (2021d) introduced adaptive stabilization and tracking control schemes designed for HOFA systems. These adaptive approaches dynamically adjust to changing system parameters, ensuring stability and performance without requiring prior knowledge of exact system characteristics. Further, Ning et al. (2022b) explored adaptive control techniques for uncertain nonlinear time-delay systems within the fully actuated framework, demonstrating their effectiveness in compensating for uncertainties and mitigating instability caused by time delays.

Recent research has also expanded HOFA system theory to enhance reliability and fault tolerance, crucial for systems operating in uncertain or hazardous environments. For instance, Cai et al. (2023a) introduced an active fault-tolerance framework for HOFA nonlinear uncertain systems, integrating adaptive observers and controllers to enable autonomous fault detection and compensation. Based on this, Cai et al. (2023c) proposed a linear observer-based fault-tolerant control strategy for time-varying fully actuated nonlinear systems, improving robustness under variable conditions. Additionally, Cai et al. (2023b) developed a fault-tolerant tracking control approach to address both multiplicative and additive actuator faults, enhancing the resilience of control systems. Despite these advancements, most existing control schemes primarily achieve asymptotic stability, where the system state gradually converges to the equilibrium over an indefinite time period. While this suffices for some practical applications, it lacks the rapid convergence required in scenarios demanding prompt responses.

Prescribed-time control has recently gained prominence due to its rapid convergence. Unlike traditional finite-time and fixed-time control methods, prescribed-time control uniquely decouples convergence time from initial states and control parameters. This enables engineers to specify a prescribed convergence time within a feasible range, providing greater flexibility across various applications. Such characteristics make prescribed-time control particularly valuable in scenarios precise settling time, such as spacecraft docking, remote healthcare interventions, and automotive emergency braking (Zhou 2020a).

The prescribed-time stabilization of high-order nonlinear systems, where nonlinearities and uncertainties are concentrated in the control input, was first explored in Song et al. (2017) using a state-scaling approach. This method employs a time-varying gain function that asymptotically approaches infinity as the settling time nears, effectively transforming state trajectories for improved convergence. Building on this foundation, numerous prescribed-time control strategies have been developed, including those in Holloway and Krstic (2019a), Holloway and Krstic (2019b), Gao et al. (2019), and Ning et al. (2022a). Specifically, Holloway and Krstic (2019a) and Holloway and Krstic (2019b) focused on prescribed-time estimation and output feedback for linear systems using observer and controllable canonical forms, respectively. Meanwhile, Gao et al. (2019) applied state-scaling transformations to stabilize switched nonlinear systems within a fixed time frame, and Ning et al. (2022a) introduced a global prescribed-time control approach for strict-feedback nonlinear systems using dynamic gain techniques to accommodate system uncertainties.

An alternative approach to prescribed-time stabilization was proposed in Krishnamurthy et al. (2020a), which introduced the temporal transformation framework as an alternative to state-scaling. In this framework, the variable $\tau \in [0, +\infty)$ maps to system time $t \in [0, T_p)$, transforming the prescribed-time stability problem in terms of t into an asymptotic stability problem relative to τ. Building on this concept, Krishnamurthy et al. (2020b) extended the framework to output feedback control for uncertain nonlinear systems. Further advancements in prescribed-time control were made by Liu and Liu (2023b), who introduced adaptive controllers for strict-feedback nonlinear systems with unknown control directions, employing a dual-transformation technique. Given the computational complexity of these approaches, particularly for high-order systems, Hua et al. (2021) developed a non-scaling backstepping method to simplify prescribed-time stability design. This approach was later extended to adaptive prescribed-time control for nonlinear systems with time delays and event-triggering mechanisms, as explored in Ning et al. (2023a), Ning et al. (2023c), and Ning et al. (2023b).

Beyond deterministic systems, stochastic prescribed-time control has been investigated for nonlinear systems under uncertainty. For instance, Li and Krstic (2022b) proposed a state-feedback control strategy, while Li and Krstic (2022a) focused on output-feedback control within a prescribed time. A notable contribution was made by Zhou (2020a), who formulated a time-varying parameter Lyapunov equation to stabilize linear systems within a prescribed time and later extended this method to nonlinear systems in Zhou and Shi (2021). However, most existing prescribed-time control methods still rely heavily on state-space modeling and first-order differential equations, often leading to complex controller designs that pose implementation challenges in practical applications.

This chapter investigates the global prescribed-time control problem for fully actuated nonlinear systems. The key contributions are summarized as follows: Unlike prescribed-time control strategies based on the state-space method, the proposed approach leverages fully actuated system theory, offering a simpler and more intuitive design methodology. A prescribed-time control strategy is developed for fully actuated nonlinear systems with unknown parameters. By utilizing the solution of the parameter Lyapunov function and adaptive techniques, this approach ensures closed-loop system stabilization and guarantees system state convergence within a specified time frame. Compared to existing control methods for fully actuated systems, which primarily achieve asymptotic stabilization, the proposed method enables prescribed-time stabilization. This ensures that convergence time remains independent of initial conditions and control parameters, allowing designers to predefine a feasible convergence period.

6.2 Time-Constrained Control for Fully Actuated Nonlinear Systems

6.2.1 System Formulation and Preliminaries

Consider a fully actuated nonlinear system defined as follows:

$$x^{(q)}(t) = f(x^{(0 \sim q-1)}(t)) + B(x^{(0 \sim q-1)}(t))u(t),$$

$$u(t) = v(t, x^{(0 \sim q-1)}(t)), \quad v(t, 0) = 0, \quad t \geq t_0, \tag{6.1}$$

where $x^{(i)}(t) \in \mathbb{R}$ ($i = 0, \ldots, q-1$) represents the system state, $t_0 \geq 0$ is the initial time, and $u(t) \in \mathbb{R}$ is the control input. The functions $f(\cdot) \in \mathbb{R}$ and $B(\cdot) \in \mathbb{R}_{>0}$ are known continuous functions satisfying $f(0) = 0$. Consequently, $x^{(0 \sim q-1)} = 0$ serves as the equilibrium point of system (6.1). Notably, the controller design relies solely on x and its derivatives.

From (6.1), one can easily get that

$$\dot{x}^{(0 \sim q-1)} = Ax^{(0 \sim q-1)} + L\left(f\left(x^{(0 \sim q-1)}\right) + Bu \right), \tag{6.2}$$

where $L = [0, 0, \ldots, 0, 1]^T \in \mathbb{R}^q$ and

$$A = \begin{bmatrix} 0 & & & \\ 0 & I_{(q-1) \times (q-1)} & \\ \vdots & & \\ 0\,0 & \cdots & 0 \end{bmatrix}. \tag{6.3}$$

Definition 6.1 For a given positive constant T_p, if there exists a control input $u \in \mathbb{R}$ ensuring that the fully actuated system (6.1) remains Lyapunov stable, and additionally satisfies

$$x^{(0 \sim q-1)}(t) = 0 \text{ for all } t \geq t_0 + T_p, \tag{6.4}$$

for any $x^{(i)}(t_0) \in \Omega$, where $i = 0, 1, \ldots, q-1$, then $x^{(0 \sim q-1)} = 0$ of system (6.1) is said to be locally prescribed-time stable. The term T_p is referred to as the prescribed time. Specifically, if $\Omega = \mathbb{R}$, $x^{(0 \sim q-1)} = 0$ is said to be globally prescribed-time stable.

Definition 6.2 (Hua et al. (2021)) Consider a continuous function $\mu(t)$. If it satisfies the following conditions:

$$\mu(t) > 0, \quad \forall t \in [t_0, t_0 + T_p),$$

$$\lim_{t \to (t_0 + T_p)^-} \int_{t_0}^{t} \mu(s) \, ds = +\infty, \tag{6.5}$$

then $\mu(t)$ is defined as a prescribed-time adjustment (T_p-PTA) function.

Lemma 6.1 (Zhou and Shi (2021)) *Consider the general parametric Lyapunov equation (PLE):*

$$A^T P(\rho) + P(\rho)A - P(\rho)LL^T P(\rho) = -\rho P(\rho). \tag{6.6}$$

This equation leads to the following properties:

(1) The PLE (6.6) has a (unique) positive definite solution $P(\rho) = W^{-1}(\rho)$, which is a polynomial function of ρ, and satisfies the Lyapunov equation

$$\left(A + \frac{\rho}{2}I_q\right) W + W \left(A + \frac{\rho}{2}I_q\right)^T = LL^T. \tag{6.7}$$

(2) The solution $P(\rho)$ exhibits the following characteristics:

$$\frac{dP(\rho)}{d\rho} > 0, \quad \text{tr}(L^T PL) = q\rho. \tag{6.8}$$

(3) There exists a constant $\delta \geq 1$, independent of ρ, ensuring:

$$\frac{dP(\rho)}{d\rho} \leq \frac{\delta P(\rho)}{q\rho}, \quad \forall \rho > 0. \tag{6.9}$$

Lemma 6.2 (Ning et al. (2023b)) *Consider the system described in (6.1) with the control input u free. Suppose there exist continuously differentiable functions $V_1(x^{(0\sim q-1)}) \in \mathbb{R}_{\geq 0}$, $V_2(t) \in \mathbb{R}_{\geq 0}$, and a class $\mathbb{K}_\infty$ function $\alpha(\cdot)$ that satisfy:*

$$V(t) = V_1(x^{(0\sim q-1)}(t)) + V_2(t), \tag{6.10}$$

$$\alpha(\|x^{(0\sim q-1)}\|) \leq V_1(x^{(0\sim q-1)}), \quad \forall x^{(i)}(t) \in \Omega, \tag{6.11}$$

$$\dot{V}(t) \leq -c\rho(t)V_1(x^{(0\sim q-1)}(t)), \quad \forall t \in [t_0, +\infty), \tag{6.12}$$

where c is a positive constant, $\rho(t) = \mu(t)$ for $t \in [t_0, t_0 + T_p)$ and $\rho(t) = \varepsilon$ for $t \in [t_0 + T_p, +\infty)$ with $\mu(t)$ as the T_p-PTA function and ε as a positive constant, then the equilibrium point of the system (6.1) is prescribed-time stable.

The objective of this chapter is to develop a control strategy for the fully actuated nonlinear systems (6.1) such that the equilibrium point of system (6.1) is prescribed-time stable in the sense of Definition 6.1.

Remark 6.1 The first-order system model in the state-space framework primarily focuses on preserving the integrity and accuracy of the system state representation. In contrast, the HOFA model is tailored for control purposes, with an emphasis on accurately determining control variables (Duan 2021a). The development of fully actuated systems is grounded in fundamental physical principles, including Newton's Law, Lagrange's Equations, etc., see Duan (2021c). Through these principles, fully actuated systems are derived by systematically eliminating underactuated components and enhancing the system to achieve greater control flexibility and precision. This transformation allows for a more comprehensive control design, facilitating the development of control strategies.

6.2.2   Time-Constrained Controller Design Based on Prescribed-Time Control

Without loss of generality, the initial condition is set to $t_0 = 0$. According to the fully actuated system theory, the controller can be designed as

$$u = -B^{-1} \left(L^T P(\rho) \left(x^{(0 \sim q-1)} \right) + f(x^{(0 \sim q-1)}) \right), t \in [0, \infty), \qquad (6.13)$$

$$\rho(t) = \begin{cases} \mu(t), & t \in [0, T_p) \\ \varepsilon, & t \in [T_p, +\infty) \end{cases}, \qquad (6.14)$$

where $\mu(t) = \frac{\sigma}{T_p - t}$ is a T_p-PTA function, and $\sigma > \frac{\delta}{q} + 1$ is a designed parameter. Then, we can obtain that

$$\dot{x}^{(0 \sim q-1)} = A \left(x^{(0 \sim q-1)} \right) - LL^T P(\rho(t)) \left(x^{(0 \sim q-1)} \right) \qquad (6.15)$$

The stability theorem is summarized as follows:

Theorem 6.1 *For the fully actuated nonlinear systems described in (6.1), under the control input given by (6.13) and (6.14), the equilibrium point $x^{(0 \sim q-1)} = 0$ of the system is globally prescribed-time stable.*

Proof First, we select the Lyapunov function V as follows:

$$V = \left(x^{(0 \sim q-1)} \right)^T P(\rho) x^{(0 \sim q-1)}. \qquad (6.16)$$

Based on the system dynamics as defined in (6.15), the derivative of the Lyapunov function V is given by:

$$\dot{V} = \left(x^{(0\sim q-1)}\right)^T \left(A^T P + PA - 2PLL^T P\right) x^{(0\sim q-1)}$$
$$+ \left(x^{(0\sim q-1)}\right)^T \left(\frac{dP}{d\rho}\dot{\rho}\right) x^{(0\sim q-1)}. \tag{6.17}$$

From (6.8), (6.9), and (6.15), we have:

$$0 \le \left(x^{(0\sim q-1)}\right)^T \left(\frac{dP}{d\rho}\dot{\rho}\right) \left(x^{(0\sim q-1)}\right) \le \frac{\delta}{\sigma q}\rho \left(x^{(0\sim q-1)}\right)^T P(\rho) \left(x^{(0\sim q-1)}\right). \tag{6.18}$$

Since $\sigma > \frac{\delta}{q} + 1$, it follows that $\frac{\delta}{\sigma q} < 1$. From (6.6), (6.17) and (6.18), we derive that:

$$\dot{V} \le -c\rho \left(x^{(0\sim q-1)}\right)^T P(\rho) \left(x^{(0\sim q-1)}\right) = -c\rho V, \tag{6.19}$$

where $c = 1 - \frac{\delta}{\sigma q}$ is a positive constant.

Then, based on (6.14) and (6.19), we deduce:

$$V(t) \le V(0) \left(1 - \frac{t}{T_p}\right)^{c\sigma}, \quad t \in [0, T_p), \tag{6.20}$$

where c is a constant with $\sigma > 1$. Let $\rho_0 = \rho(0)$, and given that:

$$V \ge \left(x^{(0\sim q-1)}\right)^T P(\rho_0) \left(x^{(0\sim q-1)}\right) \ge \lambda_{\min}(P(\rho_0)) \left\| x^{(0\sim q-1)} \right\|^2, \tag{6.21}$$

we find that:

$$\left\| x^{(0\sim q-1)}(t) \right\| \le \sqrt{\frac{V(0)}{\lambda_{\min}(P(\rho_0))}} \left(1 - \frac{t}{T_p}\right)^{c\sigma}, \quad t \in [0, T_p), \tag{6.22}$$

which imples that the state $x^{(0\sim q-1)} = 0$ converges to zero within the prescribed time.

Furthermore, combining (6.7) and (6.20), we conclude:

$$\left(x^{(0\sim q-1)}\right)^T PLL^T P \left(x^{(0\sim q-1)}\right)$$
$$\le \left(x^{(0\sim q-1)}\right)^T P^{\frac{1}{2}} \mathrm{tr}\left(P^{\frac{1}{2}} LL^T P^{\frac{1}{2}}\right) P^{\frac{1}{2}} \left(x^{(0\sim q-1)}\right)$$

$$= L^T P L Z^T P \left(x^{(0 \sim q-1)} \right) = n\rho V$$

$$\leq \frac{n V(0)}{T_p} \left(1 - \frac{t}{T_p} \right)^{\iota}, \quad t \in [0, T_p) \tag{6.23}$$

where $\iota = c\sigma - 1 > 0$ is a constant. Therefore, the controller $u(t)$ is bounded and converges to zero within the prescribed time T_p. With the aid of the continuation properties of $x^{(0 \sim q-1)}(t)$, one has:

$$\left\| x^{(0 \sim q-1)}(t) \right\| = 0, \quad u(t) = 0, \quad t \in [T_p, \infty). \tag{6.24}$$

Therefore, the equilibrium point $x^{(0 \sim q-1)} = 0$ of system (6.1) is globally prescribed-time stable. This proof is concluded. ∎

Remark 6.2 Previous studies (Duan 2021a; Cai et al. 2023b) employing fully actuated system theory have primarily achieved bounded stability or asymptotic stability, meaning that system states only converge to the origin as time approaches infinity. Unlike these approaches, the prescribed-time control strategy proposed in this chapter guarantees that system states reach the origin within a specific, pre-set time. This approach also maintains boundedness for all state variables. Importantly, the settling time of this control strategy remains constant, independent of initial system conditions, and can be configured to meet practical requirements. This feature makes it particularly suitable for applications that demand timely and predictable responses.

6.2.3 Simulation Results Validating Time-Constrained Controller

Consider the following fully actuated nonlinear systems

$$\ddot{x} = u + f(x^{(0 \sim 1)}), \tag{6.25}$$

where $f(\cdot) = 2x\dot{x} + \dot{x} \sin(x^2)$. According to the B-Section II, the controller can be designed as

$$u = -L^T P(\rho) \left(x^{(0 \sim 1)} \right) - f(\cdot), \tag{6.26}$$

with $L = [0, 1]^T$ and

$$P(\rho) = \begin{bmatrix} \rho^3 & \rho^2 \\ \rho^2 & 2\rho \end{bmatrix}, \rho(t) = \begin{cases} \frac{\sigma}{T_p - t}, & t \in [0, T_p) \\ \varepsilon, & t \in [T_p, +\infty) \end{cases}. \tag{6.27}$$

Fig. 6.1 The response of the system states $x(t)$ and $\dot{x}(t)$ under different initial conditions. (**a**) The response of state x. (**b**) The response of $\dot{x}$

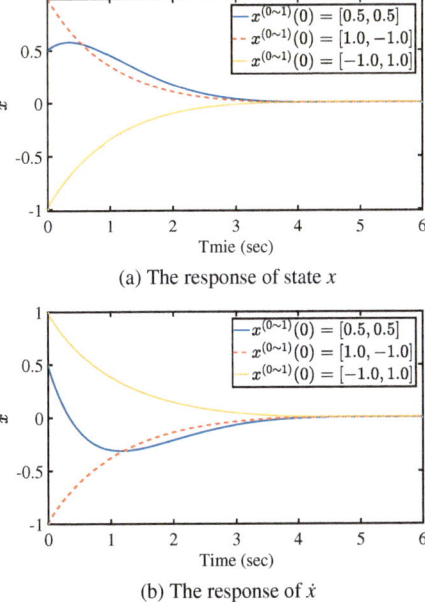

(a) The response of state x

(b) The response of $\dot{x}$

It follows from (6.27) that

$$\frac{dP(\rho)}{d\rho} \leq \frac{3P(\rho)}{\rho}. \tag{6.28}$$

Therefore, the designed parameters can be designed as

$$\sigma = 6, \quad \varepsilon = 8, \quad T_p = 5. \tag{6.29}$$

The system's initial conditions are set as

$$x^{(0 \sim q-1)}(0) = [0.5, 0.5]^T, [1, -1]^T, [-1, 1]^T.$$

The simulation results, illustrated in Figs. 6.1 and 6.2, demonstrate the responses of x and $\dot{x}$ under these conditions.

From these figures, it is evident that the x and $\dot{x}$ converge to zero within the prescribed time T_p, regardless of the initial conditions. Additionally, the controller's response, as shown in Fig. 6.2, indicates that the control signal remains bounded and also converges to zero within the prescribed time T_p across all tested initial conditions. This behavior confirms the effectiveness of the proposed control method in achieving stable, prescribed-time convergence.

Fig. 6.2 The responses of the
controller u under different
initial conditions

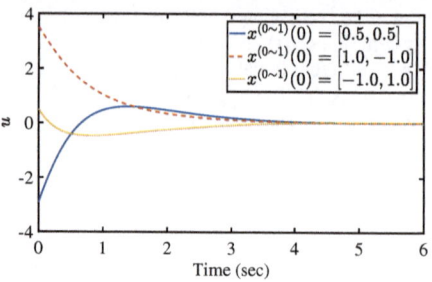

6.3   Adaptive Time-Constrained Control for Uncertain Fully Actuated Nonlinear Systems with Uncertain Parameters

6.3.1   System Formulation for Uncertain Fully Actuated Nonlinear Systems

Then fully actuated nonlinear systems with uncertain parameters are considered as follows

$$x^{(q)} = B(x^{(0 \sim q-1)})u + f(x^{(0 \sim q-1)}) + \theta^T \omega(x^{(0 \sim q-1)}), \tag{6.30}$$

where $\theta \in \mathbb{R}^r$ stands for the uncertain parameter vector and $\omega \in \mathbb{R}^r$ is a known continuous function with $\omega(0) = 0$. The system (6.30) can be transformed into

$$\dot{x}^{(0 \sim q-1)} = A\left(x^{(0 \sim q-1)}\right) + L\left(f(x^{(0 \sim q-1)}) + \theta^T \omega(x^{(0 \sim q-1)}) + Bu\right), \tag{6.31}$$

where $L = [0, 0, \ldots, 0, 1]^T \in \mathbb{R}^n$ and

$$A = \begin{bmatrix} 0 & & \\ 0 & I_{(n-1) \times (n-1)} & \\ \vdots & & \\ 0 \; 0 & \cdots & 0 \end{bmatrix}. \tag{6.32}$$

The objective of this section is to design an adaptive prescribed-time controller for the fully actuated uncertain nonlinear systems (6.30), such that the equilibrium point of system (6.30) is prescribed-time stable in the sense of Definition 6.1.

6.3.2   Adaptive Time-Constrained Controller Design

According to the fully actuated system theory, the prescribed-time controller with adaptive law can be designed as follows:

$$u(t) = -B^{-1}\left(L^T P(\rho(t))\left(x^{(0\sim q-1)}\right) + f(\cdot) + \hat{\theta}^T \omega(\cdot)\right),\tag{6.33}$$

$$\dot{\hat{\theta}}(t) = \Gamma \left(x^{(0\sim q-1)}\right)^T PL\omega,\tag{6.34}$$

$$\rho(t) = \begin{cases} \mu(t), & t \in [0, T_p) \\ \varepsilon, & t \in [T_p, +\infty) \end{cases},\tag{6.35}$$

where $\mu(t) = \frac{\sigma}{T_p - t}$ is a T_p-PTA function with the designed parameter $\sigma > \frac{\delta}{q} + 1$ and Γ is positive definite matrix.

Substituting (6.33) and (6.35) into (6.31), we have

$$\dot{x}^{(0\sim q-1)} = A\left(x^{(0\sim q-1)}\right) - LL^T P(\rho(t))\left(x^{(0\sim q-1)}\right) + L\tilde{\theta}^T \omega.\tag{6.36}$$

Theorem 6.2 *For the fully actuated nonlinear systems* (6.30) *with uncertain parameters, under the controller* (6.33)–(6.35), *then the equilibrium point of system* (6.30) *is prescribed-time stable in the sense of Definition 6.1.*

Proof First, the Lyapunov function is defined as follows:

$$V = V_1 + V_2,$$

$$V_1 = \left(x^{(0\sim q-1)}\right)^T P(\rho)\left(x^{(0\sim q-1)}\right), \quad V_2 = \tilde{\theta}^T \Gamma^{-1}\tilde{\theta}.\tag{6.37}$$

It follows from (6.36) and (6.37) that

$$\dot{V} = \left(x^{(0\sim q-1)}\right)^T \left(A^T P + P^T A - 2PLL^T P\right)\left(x^{(0\sim q-1)}\right)$$
$$+ \left(x^{(0\sim q-1)}\right)^T \left(\frac{dP}{d\rho}\dot{\rho}\right)\left(x^{(0\sim q-1)}\right)$$
$$+ 2Z^T PL\tilde{\theta}^T \omega - 2\tilde{\theta}^T \Gamma^{-1}\dot{\hat{\theta}}.\tag{6.38}$$

Based on the (6.8), (6.9) and (6.38), we can deduce that

$$0 < \left(x^{(0\sim q-1)}\right)^T \left(\frac{dP}{d\rho}\dot{\rho}\right)\left(x^{(0\sim q-1)}\right) \le \frac{\delta}{\sigma q}\rho\left(x^{(0\sim q-1)}\right)^T P(\rho)\left(x^{(0\sim q-1)}\right).$$
$$\tag{6.39}$$

Since $\sigma > \frac{\delta}{q} + 1$, we have $\frac{\delta}{\sigma q} < 1$. According to the (6.6), (6.38) and (6.39), one has

$$\dot{V} \leq -c\rho \left(x^{(0 \sim q-1)}\right)^T P(\rho) \left(x^{(0 \sim q-1)}\right) = -c\rho V_1, \tag{6.40}$$

where $c = 1 - \frac{\delta}{\sigma q} > 0$ is a constant. According to Theorem 6.1 and (6.40), one can easily get that

$$\lim_{t \to T_p^-} \| \left(x^{(0 \sim q-1)}\right) \| = \lim_{t \to T_p^-} \sqrt{\frac{V_1}{\lambda_{min}(P(\rho_0))}} = 0. \tag{6.41}$$

In addition, from (6.37) and (6.40), one has

$$\dot{V}_1 \leq -c\rho V_1 - \dot{V}_2. \tag{6.42}$$

Therefore, we can deduce that

$$V_1 \leq V_1(0) \left(1 - \frac{t}{T_p}\right)^{c\sigma} - (T_p - t)^{c\sigma} \int_0^t \frac{\dot{V}_2}{(T_p - s)^{c\sigma}} ds. \tag{6.43}$$

By taking the limits on both sides of (6.43), we have

$$\lim_{t \to T_p^-} \rho V_1 \leq \lim_{t \to T_p^-} -\frac{\sigma}{\iota} \dot{V}_2, \tag{6.44}$$

where $\iota = c\sigma - 1 > 0$ is a constant. Then, from (6.7) and (6.37), we can obtain that

$$\left(x^{(0 \sim q-1)}\right)^T PLL^T P \left(x^{(0 \sim q-1)}\right)$$
$$\leq \left(x^{(0 \sim q-1)}\right)^T P^{\frac{1}{2}} \text{tr}\left(P^{\frac{1}{2}} LL^T P^{\frac{1}{2}}\right) P^{\frac{1}{2}} \left(x^{(0 \sim q-1)}\right)$$
$$= L^T PLZ^T P \left(x^{(0 \sim q-1)}\right) = q\rho V_1. \tag{6.45}$$

According to (6.44) and (6.45), we can get that

$$\lim_{t \to T_p^-} \left(x^{(0 \sim q-1)}\right)^T PLL^T P \left(x^{(0 \sim q-1)}\right) \leq \lim_{t \to T_p^-} -\frac{\sigma}{\iota} \left(x^{(0 \sim q-1)}\right)^T PL\omega. \tag{6.46}$$

Since $\omega(0) = 0$, one has

$$\lim_{t \to T_p^-} \| \left(x^{(0 \sim q-1)}\right)^T PL \| \le \lim_{t \to T_p^-} -\frac{\sigma}{\iota}\omega(\cdot) = 0,$$

$$\| \left(x^{(0 \sim q-1)}\right)^T PL \| = 0, \ t \in [0, T_p). \tag{6.47}$$

It follows from (6.33) and (6.47) that the controller $u(t)$ is bounded and converges to zero in the prescribed-time T_p. With the aid of the continuation properties of $x^{(0 \sim q-1)}(t)$, one has:

$$\left\| x^{(0 \sim q-1)}(t) \right\| = 0, \quad u(t) = 0, \quad t \in [T_p, \infty). \tag{6.48}$$

Therefore, the equilibrium point $x^{(0 \sim q-1)} = 0$ of the system (6.30) is globally prescribed-time stable. This proof is finished. ∎

Remark 6.3 The studies in Song et al. (2017) and Gao et al. (2019) addressed the prescribed-time control problem for specific classes of nonlinear systems, while Krishnamurthy et al. (2020b) and Ning et al. (2023b) extended this investigation to uncertain nonlinear systems using adaptive techniques. all of these approaches rely on the state-space methodology for designing prescribed-time controllers, which often requires complex computational processes and poses challenges for practical implementation. This chapter proposes an approach to the adaptive prescribed-time control problem for uncertain nonlinear systems with unknown parameters by leveraging fully actuated system theory. This alternative design not only simplifies the controller structure but also enhances ease of implementation, making it a more practical choice for actual applications.

6.3.3   *Numerical Examples Highlighting Adaptive Prescribed-Time Controller*

The standard fully actuated nonlinear system with uncertain parameters is considered as follows

$$\ddot{x} = u + f(x^{(0 \sim 2)}) + \theta\omega(x^{(0 \sim 2)}), \tag{6.49}$$

where $f(\cdot) = x\sin(\dot{x})$, $\omega(\cdot) = \dot{x}\ddot{x}^2$ and $\theta = 1.2$.

Based on the B-Section, the controller can be designed as

$$u = -L^T P(\rho) \left(x^{(0\sim2)} \right) - f(\cdot) - \hat{\theta}^T \omega(\cdot),$$

$$P(\rho) = \begin{bmatrix} \rho^5 & 2\rho^4 & \rho^3 \\ 2\rho^4 & 5\rho^3 & 3\rho^2 \\ \rho^3 & 3\rho^2 & 3\rho \end{bmatrix}. \tag{6.50}$$

From (6.50), we can get that

$$\frac{dP(\rho)}{d\rho} \leq \frac{5P(\rho)}{\rho}. \tag{6.51}$$

Therefore, the design parameters can be selected as

$$\sigma = 10, \ \varepsilon = 12, \ T_p = 5 \tag{6.52}$$

In the simulation, the system is initialized with three different conditions: $x^{(0\sim q-1)}(0) = [0.5, 0.5, 0.5]^T$, $[1.0, -1.0, 1.0]^T$, and $[1.0, 1.0, -1.0]^T$. The simulation results are presented in Figs. 6.3, 6.4, and 6.5, which showing the system's responses under these initial states.

Specifically, Figs. 6.3 and 6.4a display the trajectories of x, $\dot{x}$, and $\ddot{x}$, illustrating that the system state converges to zero within the prescribed time T_p across all tested

Fig. 6.3 The response of the system states $x(t)$ and $\dot{x}(t)$ under different initial conditions. (**a**) The response of state x. (**b**) The response of $\dot{x}$

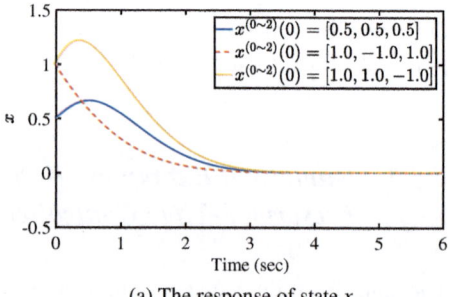

(a) The response of state x

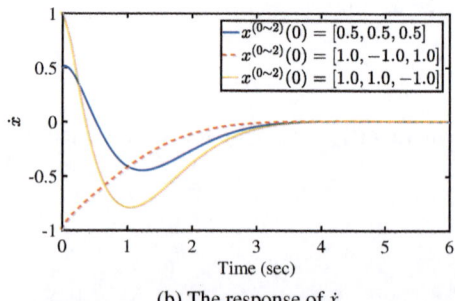

(b) The response of $\dot{x}$

Fig. 6.4 The response of the system states $\ddot{x}(t)$ and controller $u(t)$ under different initial conditions. (**a**) The response of $\ddot{x}$. (**b**) The response of u

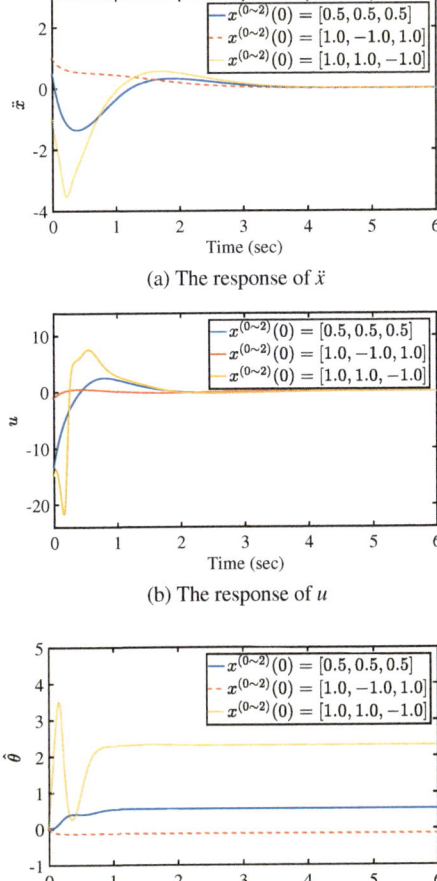

(a) The response of $\ddot{x}$

(b) The response of u

Fig. 6.5 The responses of adaptive parameter $\hat{\theta}$ under different initial conditions

conditions. Fig. 6.4b provides the controller output, confirming that the control signal remains bounded and also converges to zero within the prescribed time T_p for each initial condition. Additionally, Fig. 6.5 depicts the response of the adaptive parameter, which is shown to remain bounded throughout the process, further validating the validity of the proposed control method.

6.4 Notes and References

This chapter studies the prescribed-time control problem for fully actuated nonlinear systems. It provides a definition of prescribed-time stability for general fully actuated nonlinear systems. Drawing on the properties of the T_p-PTA function and the solution of the PLE, the chapter presents a design methodology for the prescribed-time controller tailored for fully actuated nonlinear systems. Subsequently, for fully actuated uncertain nonlinear systems, the chapter outlines the design approach for an adaptive prescribed-time controller. Utilizing the adaptive prescribed-time stability lemma, it is proved that the designed controller can make the system state converge to the origin within the predetermined time.

Part III
Adaptive State-Constrained Control for High-Order Fully Actuated Nonlinear Systems

Chapter 7
Event-Triggered Control for High-Order Fully Actuated System Under Output Constraint

This chapter addresses the design issue of event-triggered controllers for high-order fully-actuated nonlinear systems with uncertainties and output constraints. The considered system includes unknown control gain matrix and the output of the system is constrained by asymmetric time-varying functions. Based on high-order fully actuated system methodologies, an event-triggered controller is proposed, which can save the network resources used for signal transmission.

7.1 Overview

The control of nonlinear systems with output constraints has become a focal point of research, since most practical systems naturally exhibit nonlinear characteristics (Krstic et al. 1995; Li et al. 2020a) and their outputs are often limited (Yu et al. 2023; Li et al. 2022b). In recent years, Zhao and Song (2019) proposed a notable methodological method that transforms constrained systems into unconstrained ones with the same stability by using nonlinear transformation functions. Through proving the boundedness of the transformed system, this method can obtain that the original state is constrained. So far, multiple theoretical advancements have been achieved using this method. The tracking control problem for high-order uncertain nonlinear systems subject to time-varying asymmetric output constraints was considered in Ling et al. (2020). Ruan et al. (2022) proposed the adaptive fuzzy control method for uncertain multiple-input multiple-output nonlinear systems under time-varying asymmetric output constraints. For uncertain nonlinear systems with output constraints, Meng et al. (2022b) established an output feedback controller based on proportional barrier function. In conclusion, research in this domain has evolved into a comprehensive and systematic framework. It is important to note that the aforementioned studies primarily rely on first-order state-space models. While these

© The Author(s) 2026
C. Hua et al., *Adaptive Constrained Control for High-Order Fully Actuated Nonlinear Systems*, https://doi.org/10.1007/978-981-95-0962-1_7

models are better suited for analyzing state responses than for designing controllers (Duan 2021a).

HOFA systems are commonly present in practical applications (Duan 2021b), with their key characteristic being the invertibility of the control matrix. In contrast to the first-order state space models, HOFA systems preserve the full-actuation characteristics of the systems and can significantly reduce the complexity involved in controller design. Numerous significant contributions have been made regarding HOFA systems. By applying the robust control theory, the robust stabilization controllers and the robust tracking controllers for uncertain HOFA systems were constructed in Duan (2021c). Using the adaptive control theory, Duan (2021d) designed adaptive controllers for uncertain HOFA systems which can realize stabilization and tracking. Furthermore, Duan (2021e) presented the robust adaptive controllers for uncertain HOFA systems by integrating robust and adaptive control theories. The disturbance attenuation and decoupling challenges of HOFA systems were explored in Duan (2021f). For general dynamic control systems, Duan (2021g) established a controllability analysis framework and two parameter design strategies. Additionally, Duan (2022a) designed a PID controller for HOFA systems, achieving asymptotic tracking control. In the above works, the control gain matrices are completely known and can be utilized for the controller design. How to relax this condition is an interesting question.

Under the framework of continuous sampling control, the controller continuously sends signals to the actuator, regardless of whether the actuator actually requires them. To conserve network resources in signal transmission, the sampled-data method (Manivannan et al. 2018) and the event-triggered controller (Shi et al. 2020) are usually employed. Unlike the sampled-data method, the event-triggered control strategy only updates when a predefined condition is met. In early studies on event-triggered control (Tabuada 2007), the input-to-state stability (ISS) assumption was necessary to address measurement errors. Obviously, this assumption is very conservative, which prompts researchers to make efforts to eliminate this assumption. Xing et al. (2017) constructed an adaptive controller and modified the controller using the event-triggered control, which removed the ISS assumption. Song et al. (2021a) successfully applied it to the unmanned surface vehicle. For stochastic systems, Hua et al. (2018) designed an event-triggered controller based on a novel inequality, which offers a fresh approach to eliminate the ISS assumption. Li et al. (2022d) achieved adaptive compensation for measurement errors through a new transformation, thereby avoiding reliance on robust term information. For p-normal nonlinear systems with time delays, Yuan and Zhai (2022) proposed a dynamic event-triggered control algorithm with adjustable threshold parameters. Nevertheless, research on event-triggered control for HOFA systems remains limited, particularly in cases where the control matrix function is not fully accessible.

Based on the aforementioned discussion, this chapter proposes an adaptive event-triggered controller for uncertain HOFA systems with output constraints. The main contributions are outlined below

1. The developed algorithm ensures that the output remains in predefined asymmetric time-varying functions.

2. The considered control gain matrix of the uncertain HOFA system is unknown. It only requires the existence of minimum and maximum eigenvalues, which is more applicable to practical scenarios.

3. The event-triggered mechanism and its controller are designed collaboratively. It is possible to avoid the continuous update of the controller and save the network resources used for signal transmission.

7.2 System Formulation and Preliminaries

In this chapter, the following uncertain HOFA system is considered

$$
\begin{cases}
x^{(m)} = f(x^{(0 \sim m-1)}) + \Delta f(x^{(0 \sim m-1)}) + H^T(x^{(0 \sim m-1)})\theta + G(x^{(0 \sim m-1)})u \\
y = x
\end{cases}
$$
(7.1)

where $x \in \mathbb{R}^r$ is the state vector;

$$
u(t) = [u_1(t), \ldots, u_r(t)]^T \in \mathbb{R}^r,
$$

$$
y = [y_1(t), \ldots, y_r(t)]^T \in \mathbb{R}^r.
$$

are the input and output of the HOFA, respectively; $f(x^{(0 \sim m-1)}) \in \mathbb{R}^r$ denotes the known $\mathbb{C}^1$ function with $f(0_{mr}) = 0_r$; $\Delta f(x^{(0 \sim m-1)}) \in \mathbb{R}^r$ is the unknown function with $\Delta f(0_{mr}) = 0_r$; $H(x^{(0 \sim m-1)}) \in \mathbb{R}^{m \times r}$ represents the known $\mathbb{C}^1$ matrix function with $H(0_{mr}) = 0_{m \times r}$; $\theta \in \mathbb{R}^m$ is the unknown constant vector; $G(x^{(0 \sim m-1)}) \in \mathbb{R}^{r \times r}$ is the unknown control matrix function.

Remark 7.1 A variety of practical systems can be represented or converted into the formulations of HOFA systems (7.1), including RLC circuits and single-link manipulators (Duan 2021a). Although some papers have been studied for the HOFA systems (Duan 2021a,b,c), these typically assume that the control matrix $G(x^{(0 \sim m-1)})$ can be used for controller design. Besides, these works require continuous sampling and their output is unconstrained. In contrast, this chapter designs an adaptive event-triggered controller for generalized HOFA systems with unknown control gain matrices and output constraints. However, this approach complicates the design of the control algorithm. To facilitate comprehension, the architecture of the proposed adaptive event-triggered controller for (7.1) is illustrated in Fig. 7.1.

Fig. 7.1 The algorithm of adaptive event-triggered control

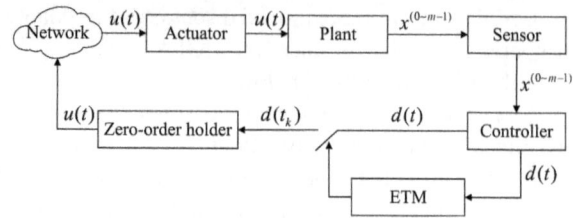

The control objective of this chapter is to design an adaptive event-triggered controller for HOFA system (7.1) such that

(i) $y_i, i = 1, \ldots, r$ are constrained in $-\underline{\varphi}_i < y_i < \overline{\varphi}_i$ for all $t \geq 0$, where $\underline{\varphi}_i, \overline{\varphi}_i$ are positive time-varying functions and their n−th derivatives are bounded and continuous.
(ii) All signals of the closed-loop system are bounded.

Some necessary Assumptions and Lemmas are given as follows.

Assumption 7.1 The system states are accessible for control design, and the initial condition of output satisfies $-\underline{\varphi}_i(0) < y_i(0) < \overline{\varphi}_i(0), i = 1, \ldots, r$.

Assumption 7.2 The control matrix $G(x^{(0\sim m-1)})$ is positive definite, and there exist unknown constants $\underline{g}$ and $\bar{g}$ such that

$$0 < \underline{g} \leq \lambda_{\min}(G(x^{(0\sim m-1)})) \leq \lambda_{\max}(G(x^{(0\sim m-1)})) \leq \bar{g} < +\infty.$$

Assumption 7.3 There are an unknown constant $s \in \mathbb{R}_+$ and a known $\mathbb{C}^1$ positive function $\varpi(x^{(0\sim m-1)}) \in \mathbb{R}$ such that $\|\Delta f(x^{(0\sim m-1)})\| \leq s\varpi(x^{(0\sim m-1)})$.

Remark 7.2 Assumptions 7.1–7.3 are conventional conditions for guaranteeing the controllability of system (7.1). The Assumption 7.2 indicates that the control gain matrix $G(x^{(0\sim m-1)})$ is bounded, which is reasonable since the system model is derived from the representation of a real-world system. The Assumption 7.3 represents a typical assumption when addressing unknown nonlinear functions. This assumption reveals that the unknown nonlinear function can be factorized into a known function and an uncertain coefficient, which retains the nonlinearity and uncertainty of $\|\Delta f(x^{(0\sim m-1)})\|$.

Lemma 7.1 (Wang et al. (2008)) *Consider a simple differential equation as*

$$\dot{a}(t) = -ka(t) + \Psi(t)$$

where k is a positive constant and $\Psi(t)$ is a nonnegative function. Then, for any $a(0) \geq 0$, $a(t) \geq 0$ holds for all $t \geq 0$.

7.3 Controller Design

7.3.1 System Transformation

To convert the initial nonlinear system (7.1) into a new unconstrained system, the subsequent system transformation is proposed

$$\xi_i = \ln\left(\frac{(\varphi_i + x_i)\overline{\varphi}_i}{(\overline{\varphi}_i - x_i)\underline{\varphi}_i}\right), i = 1, \ldots, r. \tag{7.2}$$

The time derivative of ξ_i is

$$\dot{\xi}_i = \rho_i \dot{x}_i + \tau_{i,1}, \tag{7.3}$$

where $\rho_i = \frac{\underline{\varphi}_i + \overline{\varphi}_i}{(\varphi_i + x_i)(\overline{\varphi}_i - x_i)}$ and $\tau_{i,1} = -\frac{x\dot{\varphi}_i}{\varphi_i(\varphi_i + x_i)} - \frac{x\dot{\overline{\varphi}}_i}{\overline{\varphi}_i(\overline{\varphi}_i - x_i)}$.

Take the derivative of $\xi_i^{(j)}$, $j = 2, \ldots, m$ with respect to time

$$\xi_i^{(j)} = \rho_i x_i^{(j)} + \tau_{i,j}, \tag{7.4}$$

where $\tau_{i,j} = \dot{\rho}_i x_i^{(j-1)} + \dot{\tau}_{i,j-1}$.

From (7.2), it can be obtained that

$$x_i = \frac{\overline{\varphi}_i \underline{\varphi}_i (e^{\xi_i} - 1)}{\overline{\varphi}_i + \underline{\varphi}_i e^{\xi_i}}. \tag{7.5}$$

Based on (7.4), it can be concluded that

$$x_i^{(j)} = \frac{\xi_i^{(j)} - \tau_{i,j}}{\rho_i}. \tag{7.6}$$

According to (7.1), (7.5), (7.6) and

$$\xi = [\xi_1, \xi_2, \ldots, \xi_r]^T \in \mathbb{R}^r,$$

we can get the new transformed system is

$$\xi^{(m)} = \rho(\xi^{(0\sim m-1)})\left(H^T(\xi^{(0\sim m-1)})\theta + G(\xi^{(0\sim m-1)})u + f(\xi^{(0\sim m-1)})\right)$$
$$+ \rho(\xi^{(0\sim m-1)})\Delta f(\xi^{(0\sim m-1)}) + \tau(\xi^{(0\sim m-1)}) \tag{7.7}$$

with $\rho = \mathrm{diag}(\rho_1, \rho_2, \ldots, \rho_r) \in \mathbb{R}^{r \times r}$ and

$$\tau = [\tau_{1,m}, \tau_{2,m}, \ldots, \tau_{r,m}]^T \in \mathbb{R}^r.$$

Remark 7.3 The proposed nonlinear transformation function (7.2) is smooth and strictly increasing for $-\underline{\varphi}_i < x_i < \overline{\varphi}_i, i = 1, \ldots r$ and ensures that ξ_i tends to infinity only when x_i tends to $-\underline{\varphi}_i$ or $\overline{\varphi}_i$. Therefore, the output constraints will be strictly maintained from the boundedness of ξ_i. Besides, it can be obtained from (7.5) and (7.6) that $x^{(0\sim m-1)} = 0_{mr}$ as long as $\xi^{(0\sim m-1)} = 0_{mr}$. This guarantees that it is possible to determine whether $x^{(0\sim m-1)}$ converges to zero from whether $\xi^{(0\sim m-1)}$ converges to zero.

7.3.2 Event-Triggered Mechanism Design

To address the issue of continuous controller updates and tackle the challenges posed by unknown functions, this chapter introduces the following adaptive event-triggered controller along with its ETM for $i = 1, \ldots r$ and $k \in \mathbb{N}_+$

$$d = -(1+a)r_1(\xi^{(0\sim m-1)}) - (1+a)\hat{b}r_2(\xi^{(0\sim m-1)}), \tag{7.8}$$

$$u_i = d_i(t_{i,k}); \forall t \in [t_{i,k}, t_{i,k+1}),$$

$$t_{i,k+1} = \inf\{t \in \mathbb{R}_+ | a|u_i| - |e_i| + p_i \le 0\}, \tag{7.9}$$

in which $\hat{b}$ is the estimation of $b = \frac{\bar{g}}{\underline{g}}$ and $a \in (0, 1)$ is a positive constant, where

$$d(t) = [d_1(t), \ldots, d_r(t)]^T \in \mathbb{R}^r,$$

$$e(t) = [e_1(t), \ldots, e_r(t)]^T \in \mathbb{R}^r,$$

$$p = [p_1, \ldots, p_r]^T \in \mathbb{R}^r,$$

$$t_k = [t_{1,k}, \ldots, t_{r,k}]^T \in \mathbb{R}^r,$$

and $e(t) = d(t) - u(t) \in \mathbb{R}^r$ is the measurement error, $t_{i,k}$ is the update time. r_1 and r_2 are defined as

$$r_1(\xi^{(0\sim m-1)}) = \frac{\hat{v}^2 \bar{u}(\bar{u}^T \rho(\xi^{(0\sim m-1)}) P_m^T \xi^{(0\sim m-1)})}{\sqrt{(\hat{v}^2(\xi^{(0\sim m-1)})^T P_m \rho(\xi^{(0\sim m-1)})\bar{u})^2 + \varepsilon_1^2}} \in \mathbb{R}^r,$$

$$r_2(\xi^{(0\sim m-1)}) = \frac{\bar{p}(\bar{p}^T \rho(\xi^{(0\sim m-1)}) P_m^T \xi^{(0\sim m-1)})}{\sqrt{((\xi^{(0\sim m-1)})^T P_m \rho(\xi^{(0\sim m-1)})\bar{p})^2 + \varepsilon_1^2}} \in \mathbb{R}^r,$$

where $\varepsilon_1 \in \mathbb{R}_+$, $\bar{p} = \frac{p}{1-a} \in \mathbb{R}^r$, $\hat{v}$ is the estimation of $v = \frac{1}{g}$ and $\bar{u} \in \mathbb{R}^r$ is the virtual controller in (7.10).

7.3.3 Adaptive Event-Triggered Controller Design

Taking into account the unknown function $G(x^{(0\sim m-1)})$, $\bar{u}$ is designed as

$$\bar{u} = \frac{A^{0\sim m-1}\xi^{(0\sim m-1)} + \tau(\xi^{(0\sim m-1)})}{\rho(\xi^{(0\sim m-1)})} + f(\xi^{(0\sim m-1)}) + \Psi^T(\xi^{(0\sim m-1)})\hat{\Theta},$$

(7.10)

with

$$\Psi(\xi^{(0\sim m-1)}) = [H^T(\xi^{(0\sim m-1)}), \zeta(\xi^{(0\sim m-1)})]^T \in \mathbb{R}^{(m+1)\times r},$$

$$\zeta(\xi^{(0\sim m-1)}) = \frac{\varpi^2 \rho(\xi^{(0\sim m-1)}) P_m^T \xi^{(0\sim m-1)}}{\sqrt{\varpi^2 \|(\xi^{(0\sim m-1)})^T P_m \rho(\xi^{(0\sim m-1)})\|^2 + \varepsilon_2^2}} \in \mathbb{R}^r,$$

where $\varepsilon_2 \in \mathbb{R}_+$ and $\hat{\Theta} = [\hat{\theta}, \hat{s}]^T \in \mathbb{R}^{m+1}$ is the estimation of $\Theta = [\theta, s]^T \in \mathbb{R}^{m+1}$. From (7.7) and (7.10), it can be obtained that

$$\xi^{(m)} + A^{0\sim m-1}\xi^{(0\sim m-1)} = \phi(\xi^{(0\sim m-1)}),$$

(7.11)

where

$$\phi(\xi^{(0\sim m-1)}) = \rho(\xi^{(0\sim m-1)})\Delta f(\xi^{(0\sim m-1)}) + \rho(\xi^{(0\sim m-1)})H^T(\xi^{(0\sim m-1)})\tilde{\theta}$$
$$- \rho(\xi^{(0\sim m-1)})\zeta(\xi^{(0\sim m-1)})\hat{s} + \rho(\xi^{(0\sim m-1)})G(\xi^{(0\sim m-1)})u$$
$$+ \rho(\xi^{(0\sim m-1)})\bar{u} \in \mathbb{R}^r.$$

Furthermore, the HOFA system (7.11) can be rewritten as

$$\dot{\xi}^{(0\sim m-1)} = \Phi(A^{0\sim m-1})\xi^{(0\sim m-1)} + \begin{bmatrix} 0_{(m-1)r} \\ \phi(\xi^{(0\sim m-1)}) \end{bmatrix} \in \mathbb{R}^{mr}.$$

(7.12)

where $A_i \in \mathbb{R}^{r\times r}$ in $\Phi(A^{0\sim m-1})$ satisfies the condition in Lemma 2.6. Therefore, the following Lyapunov function is selected

$$V = \frac{1}{2}(\xi^{(0\sim m-1)})^T P \xi^{(0\sim m-1)} + \frac{g\tilde{v}^2}{2} + \frac{1}{2}\tilde{\Theta}^T \Gamma^{-1}\tilde{\Theta} + \frac{g\tilde{b}^2}{2},$$

(7.13)

where $\tilde{v}(t) = v - \hat{v}(t) \in \mathbb{R}$, $\tilde{\Theta}(t) = \Theta - \hat{\Theta}(t) \in \mathbb{R}^{m+1}$, $\tilde{b}(t) = b - \hat{b}(t) \in \mathbb{R}$ and $\Gamma \in \mathbb{R}^{(m+1)\times(m+1)}$ is a positive definite matrix. Based on Lemma 2.6, the time derivative of (7.13) is

$$
\begin{aligned}
\dot{V} = {} & \frac{1}{2}\left(\xi^{(0\sim m-1)}\right)^T (P\Phi + \Phi^T P)\xi^{(0\sim m-1)} + \left(\xi^{(0\sim m-1)}\right)^T P \begin{bmatrix} 0_{(m-1)r} \\ \phi(\xi^{(0\sim m-1)}) \end{bmatrix} \\
& - \tilde{\Theta}^T \Gamma^{-1}\dot{\hat{\Theta}} - \underline{g}\tilde{v}\dot{\hat{v}} - \underline{g}\tilde{b}\dot{\hat{b}} \\
\leq {} & \left(\xi^{(0\sim m-1)}\right)^T P_m \rho(\xi^{(0\sim m-1)})\left(\Delta f(\xi^{(0\sim m-1)}) + H^T(\xi^{(0\sim m-1)})\tilde{\theta} + \bar{u} \right. \\
& \left. - \zeta(\xi^{(0\sim m-1)})\hat{s} + G(\xi^{(0\sim m-1)})u\right) - \frac{\mu}{2}\left(\xi^{(0\sim m-1)}\right)^T P\xi^{(0\sim m-1)} \\
& - \tilde{\Theta}^T \Gamma^{-1}\dot{\hat{\Theta}} - \underline{g}\tilde{v}\dot{\hat{v}} - \underline{g}\tilde{b}\dot{\hat{b}}.
\end{aligned}
\tag{7.14}
$$

It can be concluded from Assumption 7.2 and Lemma 2.8 that

$$
\begin{aligned}
& \left(\xi^{(0\sim m-1)}\right)^T P_m \rho(\xi^{(0\sim m-1)})\Delta f(\xi^{(0\sim m-1)}) \\
& \leq s\left(\xi^{(0\sim m-1)}\right)^T P_m \rho(\xi^{(0\sim m-1)})\zeta(\xi^{(0\sim m-1)}) + s\varepsilon_2.
\end{aligned}
\tag{7.15}
$$

Substituting (7.15) into (7.14), it follows that

$$
\begin{aligned}
\dot{V} \leq {} & -\frac{\mu}{2}\left(\xi^{(0\sim m-1)}\right)^T P\xi^{(0\sim m-1)} + \left(\xi^{(0\sim m-1)}\right)^T P_m \left(\rho(\xi^{(0\sim m-1)})\bar{u} \right. \\
& \left. + \rho(\xi^{(0\sim m-1)})\Psi^T(\xi^{(0\sim m-1)})\tilde{\Theta} + \rho(\xi^{(0\sim m-1)})G(x^{(0\sim m-1)})u\right) \\
& - \tilde{\Theta}^T \Gamma^{-1}\dot{\hat{\Theta}} - \underline{g}\tilde{b}\dot{\hat{b}} + s\varepsilon_2 - \underline{g}\tilde{v}\dot{\hat{v}}.
\end{aligned}
\tag{7.16}
$$

From (7.8), we have

$$
d_i(t) = (1 + \kappa_1(t)a)u_i(t) + \kappa_2(t)p_i,
\tag{7.17}
$$

where $\kappa_1(t)$ and $\kappa_2(t)$ are time-varying parameters satisfying $|\kappa_1(t)| \leq 1$ and $|\kappa_2(t)| \leq 1$.

It can be known from (7.8) that

$$
d(t) = (1 + \kappa_1(t)a)u(t) + \kappa_2(t)p.
\tag{7.18}
$$

Then, it can be obtained

$$u = \frac{d}{1 + \kappa_1 a} - \frac{\kappa_2 p}{1 + \kappa_1 a}. \tag{7.19}$$

It follows from (7.8) that

$$\left(\xi^{(0 \sim m-1)}\right)^T P_m \rho(\xi^{(0 \sim m-1)}) G(\xi^{(0 \sim m-1)}) d \leq 0. \tag{7.20}$$

Combined with (7.20) and Lemma 2.8, it can be acquired that

$$
\begin{aligned}
&\left(\xi^{(0 \sim m-1)}\right)^T P_m \rho(\xi^{(0 \sim m-1)}) G(\xi^{(0 \sim m-1)}) u \\
&= \frac{\left(\xi^{(0 \sim m-1)}\right)^T P_m \rho(\xi^{(0 \sim m-1)}) G(\xi^{(0 \sim m-1)}) (d - \kappa_2 p)}{1 + \kappa_1 a} \\
&\leq \bar{g} \left| \frac{\left(\xi^{(0 \sim m-1)}\right)^T P_m \rho(\xi^{(0 \sim m-1)}) p}{1 - a} \right| \\
&\quad + \underline{g} \frac{\left(\xi^{(0 \sim m-1)}\right)^T P_m \rho(\xi^{(0 \sim m-1)}) d}{1 + a} \\
&\leq \bar{g} (\xi^{(0 \sim m-1)})^T P_m \rho(\xi^{(0 \sim m-1)}) r_2(\xi^{(0 \sim m-1)}) \\
&\quad + \underline{g} \frac{\left(\xi^{(0 \sim m-1)}\right)^T P_m \rho(\xi^{(0 \sim m-1)}) d}{1 + a} + \varepsilon_1 \bar{g}.
\end{aligned} \tag{7.21}
$$

It follows from (7.8) and (7.21) that

$$
\begin{aligned}
&\underline{g} \frac{\left(\xi^{(0 \sim m-1)}\right)^T P_m \rho(\xi^{(0 \sim m-1)}) d}{1 + a} \\
&\leq -\underline{g} \hat{v}(\xi^{(0 \sim m-1)})^T P_m \rho(\xi^{(0 \sim m-1)}) \bar{u} + \varepsilon_1 \underline{g} - \underline{g} \hat{b}(\xi^{(0 \sim m-1)})^T P_m \rho(\xi^{(0 \sim m-1)}) r_2.
\end{aligned} \tag{7.22}
$$

Taking (7.22) into (7.21), we can derive that

$$
\begin{aligned}
&\left(\xi^{(0 \sim m-1)}\right)^T P_m \rho(\xi^{(0 \sim m-1)}) G(\xi^{(0 \sim m-1)}) u + \underline{g} v(\xi^{(0 \sim m-1)})^T P_m \rho(\xi^{(0 \sim m-1)}) \bar{u} \\
&\leq \underline{g} \tilde{v}(\xi^{(0 \sim m-1)})^T P_m \rho(\xi^{(0 \sim m-1)}) \bar{u} + \varepsilon_1 \underline{g} \\
&\quad + \varepsilon_1 \bar{g} + \underline{g} \tilde{b}(\xi^{(0 \sim m-1)})^T P_m \rho(\xi^{(0 \sim m-1)}) r_2.
\end{aligned} \tag{7.23}
$$

Bringing (7.23) into (7.16), it shows that

$$
\begin{aligned}
\dot{V} \leq & -\frac{\mu}{2}\left(\xi^{(0\sim m-1)}\right)^T P \xi^{(0\sim m-1)} + \varepsilon_1 \underline{g} + \varepsilon_1 \bar{g} + s\varepsilon_2 \\
& - \tilde{\Theta}^T \Gamma^{-1}(\dot{\hat{\Theta}} - \Gamma \Psi \rho(\xi^{(0\sim m-1)}) P_m^T \xi^{(0\sim m-1)}) \\
& - g\tilde{b}(\dot{\hat{b}} - (\xi^{(0\sim m-1)})^T P_m \rho(\xi^{(0\sim m-1)}) r_2) \\
& - g\tilde{v}(\dot{\hat{v}} - (\xi^{(0\sim m-1)})^T P_m \rho(\xi^{(0\sim m-1)}) \bar{u}).
\end{aligned} \tag{7.24}
$$

The adaptive laws are designed as

$$
\begin{aligned}
\dot{\hat{v}} &= (\xi^{(0\sim m-1)})^T P_m \rho(\xi^{(0\sim m-1)}) \bar{u} - \sigma_1 \hat{v}, \\
\dot{\hat{\Theta}} &= \Gamma \Psi \rho(\xi^{(0\sim m-1)}) P_m^T \xi^{(0\sim m-1)} - \sigma_2 \Gamma \hat{\Theta}, \\
\dot{\hat{b}} &= (\xi^{(0\sim m-1)})^T P_m \rho(\xi^{(0\sim m-1)}) r_2 - \sigma_3 \hat{b}.
\end{aligned} \tag{7.25}
$$

where $\sigma_1, \sigma_2, \sigma_3$ are positive constants.
Substituting (7.25) into (7.24), it can be acquired that

$$
\begin{aligned}
\dot{V} \leq & -\frac{\mu}{2}\left(\xi^{(0\sim m-1)}\right)^T P \xi^{(0\sim m-1)} - \frac{\sigma_1 g}{2}\tilde{v}^2 - \frac{\sigma_2}{2}\tilde{\Theta}^T \tilde{\Theta} - \frac{\sigma_3 g}{2}\tilde{b}^2 \\
& + \frac{\sigma_1 g}{2}v^2 + \frac{\sigma_3 g}{2}b^2 + \frac{\sigma_2}{2}\Theta^T \Theta + \varepsilon_1 \underline{g} + \varepsilon_1 \bar{g} + s\varepsilon_2 \\
\leq & -cV + q,
\end{aligned} \tag{7.26}
$$

where

$$
c = \min\{\mu, \sigma_1, \sigma_2/\lambda_{\max}(\Gamma^{-1}), \sigma_3\},
$$

$$
q = \frac{\sigma_1 g}{2}v^2 + \varepsilon_1 \underline{g} + \frac{\sigma_2}{2}\Theta^T \Theta + \varepsilon_1 \bar{g} + \frac{\sigma_3 g}{2}b^2 + s\varepsilon_2.
$$

Theorem 7.1 *For the uncertain HOFA system (7.1) satisfying Assumptions 7.1–7.3, the adaptive event-triggered controller (7.8) can ensure that all signals of the closed-loop system are bounded and* $-\underline{\varphi}_i < x_i < \bar{\varphi}_i, i = 1, ..., r.$

Proof It is straightforward to obtain from (7.26) that

$$
V \leq e^{-ct}(V(0) - \frac{q}{c}) + \frac{q}{c}. \tag{7.27}
$$

which means that, when $t \to +\infty$,

$$V \to \frac{q}{c}. \tag{7.28}$$

Such that $\xi^{(j)}$, $j = 0, \ldots, m - 1$, $\hat{\Theta}$, $\hat{v}$, $\hat{b}$, $\varpi(\xi^{(0 \sim m-1)})$, $\Psi(\xi^{(0 \sim m-1)})$ are bounded. From (7.8)–(7.10), the boundedness of d, u and $\bar{u}$ can be concluded. Combined with (7.5) and (7.6), it can be known that $x^{(j)}$ are bounded. Therefore, all signals of the closed-loop system are bounded. Besides, it can be derived from (7.2) that the boundedness of ξ_i implies $-\underline{\varphi}_i < x_i < \overline{\varphi}_i$.

The rest of the work is to demonstrate that the proposed ETM does not exhibit Zeno behavior. For $k \in \mathbb{N}_+$, suppose that there is a time interval $\underline{t} = [\underline{t}_1, \ldots, \underline{t}_r]^T \in \mathbb{R}^r$ satisfying

$$\underline{t} \leq t_{k+1} - t_k. \tag{7.29}$$

From $e_i(t) = d_i(t) - u_i(t)$ for $t \in [t_{i,k}, t_{i,k+1})$, it can be get

$$\frac{d|e_i|}{dt} = \text{sign}(e_i)\dot{e}_i \leq |\dot{d}_i|. \tag{7.30}$$

It is important to highlight that $\varpi(\xi^{(0 \sim m-1)})$ and $H(\xi^{(0 \sim m-1)})$ possess the first-order continuous partial derivatives, it is straightforward to conclude that $\dot{d}$ is continuous. Due to the boundedness of all signals of the closed-loop, there must be a constant $\varsigma_i \in \mathbb{R}_+$ such that

$$|\dot{d}_i| \leq \varsigma_i. \tag{7.31}$$

From (7.9), it can be obtained in that $e_i(t_{i,k}) = 0$ and $e_i(t_{i,k+1}) = a|u_i| + p_i$ hold. These derive that

$$\underline{t}_i \geq \frac{a|u_i| + p_i}{\varsigma_i} \geq \frac{p_i}{\varsigma_i}. \tag{7.32}$$

Therefore, the time interval $\underline{t}$ satisfies

$$\underline{t} \geq \omega, \tag{7.33}$$

where $\omega \in \mathbb{R}_+$. Namely, the proposed ETM is Zeno-free.

The proof is complete.

To present the workflow of the proposed control algorithm more clearly, the following table is presented

Remark 7.4 Although Ruan et al. (2022) considered the incompletely known gain function, it designed controller based on the backstepping method, which will lead to computational explosion. To handle the problem, we construct an adaptive

---

Algorithm of adaptive event-triggered control

Initialization:

(i) Check whether the HOFA system (7.1) complies with Assumptions 7.2–7.3.

(ii) Set the initial values: $x^{(0\sim m-1)}(0)$, $\hat{\Theta}(0)$, $\hat{v}(0)$, $\hat{b}(0)$;

(iii) Determine the output constraints boundaries and check Assumption 7.1.

(iv) Based on Lemma 2.6, calculate the matrix P once given $A_0, \ldots, A_{m-1}$;

While $t < T$, where T denotes the lifespan of the HOFA system (7.1)

　　Do　Calculate $d_i(t)$ and the measurement error $e_i(t)$

　　　　For $i = \mathbb{N}_{1:r}$

　　　　If The ETM (7.9) is satisfied

　　　　　　(i) Set $t_{i,k+1} = t$, $u_i(t) = u_i(t_{i,k+1})$, and $e_i(t) = 0$;

　　　　　　(ii) Send $u_i(t)$ to actuator.

　　　　Else $u_i(t) = u_i(t_{i,k})$, and $e_i(t) = d_i(t) - u_i(t_{i,k})$.

End All

---

event-triggered controller based on the HOFA system, whose controller structure is simpler and retains the original physical background of the systems. In Duan (2021a,b,c,d,e), only the completely known control matrix functions are taken into account. Different from these works, we allow that the control matrix function is positive definite and has the unknown maximum and minimum eigenvalues, which causes that the existing control algorithms are not effective for the challenging problem we have concerned in this chapter.

Remark 7.5 In this chapter, the biggest difficulty is caused by the unknown control matrix function. Considering the unknown control matrix function, the virtual controller $\bar{u}$ is introduced in (7.10), and the adaptive law $\hat{v}$ is designed to estimate $v = \frac{1}{g}$. Based on the ETM (7.9), it can be obtained that $u = \frac{d}{1+o_1 a} - \frac{o_2 p}{1+o_1 a}$. To ensure that $(\xi^{(0\sim m-1)})^T P_m \rho(\xi^{(0\sim m-1)})\bar{u}$ in (7.14) can be offset on the premise that $(\xi^{(0\sim m-1)})^T P_m \rho(\xi^{(0\sim m-1)})G(\xi^{(0\sim m-1)})d$ is positive, $-(1+a)r_1(\xi^{(0\sim m-1)})$ is constructed in (7.10). In addition, due to the existence of the constant p in ETM, $\bar{g}(\xi^{(0\sim m-1)})^T P_m \rho(\xi^{(0\sim m-1)})r_2$ appears in (7.23). To deal with the term, we introduce an adaptive law $\hat{b}$ to transform the coefficient $\bar{g}$ into the coefficient $\underline{g}$, which enables the term to be compensated directly in the controller, namely, the term is rewritten as $\underline{g}b(\xi^{(0\sim m-1)})^T P_m \rho(\xi^{(0\sim m-1)})r_2$. Then, by establishing $-(1+a)\hat{b}r_2(\xi^{(0\sim m-1)})$ in (7.10), the effect caused by the constant p is successfully overcome.

Remark 7.6 The controller proposed in this chapter can guarantee that $\xi^{(0\sim m-1)}$, $x^{(0\sim m-1)}$, $\tilde{\Theta}$, $\tilde{v}$, $\tilde{b}$, d tend to 0 when $t \to +\infty$ by taking $\sigma_1 = \sigma_2 = \sigma_3 = 0$, $\varepsilon_1 = e^{-\gamma_1 t}$ and $\varepsilon_2 = e^{-\gamma_2 t}$ with $\gamma_1 \in \mathbb{R}_+$ and $\gamma_2 \in \mathbb{R}_+$. In this case, (7.26) can be rewritten as $\dot{V} \leq -cV + q_0 \leq q_0$, where $q_0 = \varepsilon_1 \underline{g} + \varepsilon_1 \bar{g} + s\varepsilon_2$. From the values of ε_1 and ε_2, there is a constant $\bar{\varepsilon}_i \in \mathbb{R}_+, i = 1, 2$ such that $\bar{\varepsilon}_i \geq \int_0^t \varepsilon_i(\iota)d\iota$. Then, it can be calculated that $V(t) \leq V(0) + \varepsilon$ with $\varepsilon = \bar{\varepsilon}_1 \underline{g} + \bar{\varepsilon}_1 \bar{g} + s\bar{\varepsilon}_2$. Therefore, the boundedness of $\xi^{(0\sim m-1)}$, $x^{(0\sim m-1)}$, $\tilde{\Theta}$, $\tilde{v}$, $\tilde{b}$ can be obtained on $t \in [0, t_f)$. Moreover, there is no phenomenon of finite-time escape at $t_f = \infty$. Therefore, $\xi^{(0\sim m-1)}$, $x^{(0\sim m-1)}$, $\tilde{\Theta}$, $\tilde{v}$, $\tilde{b}$ are square integrable and their time derivatives are bounded. It follows from the Barbalat's Lemma that $\xi^{(0\sim m-1)}$, $x^{(0\sim m-1)}$, $\tilde{\Theta}$, $\tilde{v}$, $\tilde{b}$ converge to 0 when $t \to \infty$. Then $\varpi(\xi^{(0\sim m-1)})$ and $H(\xi^{(0\sim m-1)})$ will converge to 0 when $t \to \infty$. From (7.8), d will also tend to 0 when $t \to \infty$. Therefore, $\xi^{(0\sim m-1)}$, $x^{(0\sim m-1)}$, $\tilde{\Theta}$, $\tilde{v}$, $\tilde{b}$, d converge to zero when $t \to \infty$.

7.3.4 Two Parameter Estimation Based on Event-Triggered Control Design

The unknown parameters in the system (7.1) pose challenges for controller design, and the increase in the number of adaptive parameters also leads to an increase in computational burden. In this chapter, the number of adaptive parameters will be reduced.

For $i = 1, \ldots r$ and $k \in \mathbb{N}_+$, the adaptive event-triggered controller along with its ETM is changed to

$$\check{d} = -(1+a)\check{r}_1(\xi^{(0\sim m-1)}) - (1+a)\hat{e}r_2(\xi^{(0\sim m-1)}), \tag{7.34}$$

$$u_i = \check{d}_i(\check{t}_{i,k}); \forall t \in [\check{t}_{i,k}, \check{t}_{i,k+1}), , $$

$$\check{t}_{i,k+1} = \inf\{t \in \mathbb{R}_+ | a|u_i| - |\check{e}_i| + p_i \le 0\}, \tag{7.35}$$

where $\check{d}(t) = [\check{d}_1(t), \ldots, \check{d}_r(t)]^T \in \mathbb{R}^r$, $\check{e}(t) = [\check{e}_1(t), \ldots, \check{e}_r(t)]^T \in \mathbb{R}^r$, $\check{e}(t) = \check{d}(t) - u(t) \in \mathbb{R}^r$, $p = [p_1, \ldots, p_r]^T \in \mathbb{R}^r$, $\check{t}_k = [\check{t}_{1,k}, \ldots, \check{t}_{r,k}]^T \in \mathbb{R}^r$ and $\check{t}_{i,k}$ is the update time.

$\check{r}_1$ and r_2 are defined as

$$\check{r}_1(\xi^{(0\sim m-1)}) = \frac{\hat{v}^2 \check{\bar{u}}(\check{\bar{u}}^T \rho(\xi^{(0\sim m-1)}) P_m^T \xi^{(0\sim m-1)})}{\sqrt{(\hat{v}^2(\xi^{(0\sim m-1)})^T P_m \rho(\xi^{(0\sim m-1)})\check{\bar{u}})^2 + \varepsilon_1^2}} \in \mathbb{R}^r,$$

$$r_2(\xi^{(0\sim m-1)}) = \frac{\bar{p}(\bar{p}^T \rho(\xi^{(0\sim m-1)}) P_m^T \xi^{(0\sim m-1)})}{\sqrt{((\xi^{(0\sim m-1)})^T P_m \rho(\xi^{(0\sim m-1)})\bar{p})^2 + \varepsilon_1^2}} \in \mathbb{R}^r,$$

in which $\varepsilon_1 \in \mathbb{R}_+$, $\bar{p} = \frac{p}{1-a} \in \mathbb{R}^r$, $\hat{v}$ is the estimation of $v = \frac{1}{g}$.

$\check{u} \in \mathbb{R}^r$ is designed as

$$
\begin{aligned}
\check{u} = &\frac{A^{0 \sim m-1} \xi^{(0 \sim m-1)} + \tau(\xi^{(0 \sim m-1)})}{\rho(\xi^{(0 \sim m-1)})} + f(\xi^{(0 \sim m-1)}) \\
&+ (\zeta(\xi^{(0 \sim m-1)}) + \delta(\xi^{(0 \sim m-1)}))\hat{\epsilon},
\end{aligned}
\tag{7.36}
$$

where

$$
\zeta(\xi^{(0 \sim m-1)}) = \frac{\varpi^2 \rho(\xi^{(0 \sim m-1)}) P_m^T \xi^{(0 \sim m-1)}}{\sqrt{\varpi^2 \|(\xi^{(0 \sim m-1)})^T P_m \rho(\xi^{(0 \sim m-1)})\|^2 + \varepsilon_2^2}} \in \mathbb{R}^r,
$$

$$
\delta(\xi^{(0 \sim m-1)}) = \frac{H^T(\xi^{(0 \sim m-1)}) H(\xi^{(0 \sim m-1)}) \rho(\xi^{(0 \sim m-1)}) P_m^T \xi^{(0 \sim m-1)}}{\sqrt{\|(\xi^{(0 \sim m-1)})^T P_m \rho(\xi^{(0 \sim m-1)}) H^T(\xi^{(0 \sim m-1)})\|^2 + \varepsilon_2^2}} \in \mathbb{R}^r,
$$

$\hat{\epsilon}$ is the estimate of $\epsilon = \max\{\frac{\|\theta\|}{\eta}, \frac{s}{\eta}, \frac{\bar{g}}{\eta}\}$, $\eta = \min\{1, \underline{g}\}$ and $\varepsilon_2 \in \mathbb{R}_+$.
    The adaptive laws of $\hat{v}$ and $\hat{\epsilon}$ are designed as

$$
\begin{aligned}
\dot{\hat{v}} = &(\xi^{(0 \sim m-1)})^T P_m \rho(\xi^{(0 \sim m-1)})\check{u} - \sigma_1 \hat{v}, \\
\dot{\hat{\epsilon}} = &(\xi^{(0 \sim m-1)})^T P_m \Big(\rho(\xi^{(0 \sim m-1)})(\zeta(\xi^{(0 \sim m-1)}) \\
&+ \delta(\xi^{(0 \sim m-1)})) + \rho(\xi^{(0 \sim m-1)}) r_2 \Big) - \sigma_2 \hat{\epsilon},
\end{aligned}
\tag{7.37}
$$

where σ_1 and σ_2 are positive constants. The initial values of $\hat{\rho}$ and $\hat{\epsilon}$ must be non-negative numbers, that is, $\sigma_1(0) \geq 0$ and $\sigma_2(0) \geq 0$.
    The following theorem summarizes the main results of this chapter.

Theorem 7.2 *For the uncertain HOFA system (7.1) satisfying Assumptions 7.1–7.3, the adaptive event-triggered controller (7.34) can ensure that*

 (i) all signals of the closed-loop system are bounded and
(ii) $-\underline{\varphi}_i < x_i < \overline{\varphi}_i, i = 1, ..., r$.

Proof Combined with (7.36) and (7.7) can be rewritten as

$$
\xi^{(m)} + A^{0 \sim m-1} \xi^{(0 \sim m-1)} = \check{\phi}(\xi^{(0 \sim m-1)}),
\tag{7.38}
$$

where

$$
\begin{aligned}
\check{\phi}(\xi^{(0 \sim m-1)}) = &\rho(\xi^{(0 \sim m-1)}) \Delta f(\xi^{(0 \sim m-1)}) + \rho(\xi^{(0 \sim m-1)}) H^T(\xi^{(0 \sim m-1)})\theta \\
&- \rho(\xi^{(0 \sim m-1)})(\zeta(\xi^{(0 \sim m-1)}) + \delta(\xi^{(0 \sim m-1)}))\hat{\epsilon} \\
&+ \rho(\xi^{(0 \sim m-1)}) G(\xi^{(0 \sim m-1)}) u + \rho(\xi^{(0 \sim m-1)})\check{u} \in \mathbb{R}^r.
\end{aligned}
$$

The HOFA system (7.38) is equals to

$$\dot{\xi}^{(0\sim m-1)} = \Phi(A^{0\sim m-1})\xi^{(0\sim m-1)} + \begin{bmatrix} 0_{(m-1)r} \\ \breve{\phi}(\xi^{(0\sim m-1)}) \end{bmatrix} \in \mathbb{R}^{mr}. \tag{7.39}$$

where $A_i \in \mathbb{R}^{r \times r}$ in $\Phi(A^{0\sim m-1})$ satisfies the condition in Lemma 2.6. Therefore, there exists a positive definite matrix $P(A^{0\sim m-1})$ satisfying Lemma 2.6.

In this chapter, the following Lyapunov function is selected

$$V = \frac{1}{2}\left(\xi^{(0\sim m-1)}\right)^T P\xi^{(0\sim m-1)} + \frac{g\tilde{v}^2}{2} + \frac{\eta}{2}\tilde{\epsilon}^2, \tag{7.40}$$

where $\tilde{v}(t) = v - \hat{v}(t) \in \mathbb{R}$, $\tilde{\epsilon}(t) = \epsilon - \hat{\epsilon}(t) \in \mathbb{R}$. Based on Lemma 2.6, the time derivative of (7.40) is

$$\begin{aligned}
\dot{V} =& \frac{1}{2}\left(\xi^{(0\sim m-1)}\right)^T (P\Phi + \Phi^T P)\xi^{(0\sim m-1)} \\
&+ \left(\xi^{(0\sim m-1)}\right)^T P \begin{bmatrix} 0_{(m-1)r} \\ \breve{\phi}(\xi^{(0\sim m-1)}) \end{bmatrix} - g\tilde{v}\dot{\hat{v}} - \eta\tilde{\epsilon}\dot{\hat{\epsilon}} \\
=& -\frac{\mu}{2}\left(\xi^{(0\sim m-1)}\right)^T P\xi^{(0\sim m-1)} - g\tilde{v}\dot{\hat{v}} - \eta\tilde{\epsilon}\dot{\hat{\epsilon}} \\
&+ \left(\xi^{(0\sim m-1)}\right)^T P_m\left(\rho(\xi^{(0\sim m-1)})\Delta f(\xi^{(0\sim m-1)})\right. \\
&+ \rho(\xi^{(0\sim m-1)})H^T(\xi^{(0\sim m-1)})\theta \\
&- \rho(\xi^{(0\sim m-1)})(\zeta(\xi^{(0\sim m-1)}) + \delta(\xi^{(0\sim m-1)}))\hat{\epsilon} \\
&+ \rho(\xi^{(0\sim m-1)})G(\xi^{(0\sim m-1)})u \\
&+ \left.\rho(\xi^{(0\sim m-1)})\breve{u}\right).
\end{aligned} \tag{7.41}$$

Upon using Assumption 7.2 and Lemma 2.8, we have

$$\begin{aligned}
&\left(\xi^{(0\sim m-1)}\right)^T P_m\rho(\xi^{(0\sim m-1)})\Delta f(\xi^{(0\sim m-1)}) \\
&\leq s\varpi\left\|\left(\xi^{(0\sim m-1)}\right)^T P_m\rho(\xi^{(0\sim m-1)})\right\| \\
&\leq s\left(\xi^{(0\sim m-1)}\right)^T P_m\rho(\xi^{(0\sim m-1)})\zeta(\xi^{(0\sim m-1)}) + s\varepsilon_2 \\
&\leq \eta\epsilon\left(\xi^{(0\sim m-1)}\right)^T P_m\rho(\xi^{(0\sim m-1)})\zeta(\xi^{(0\sim m-1)}) + s\varepsilon_2.
\end{aligned} \tag{7.42}$$

$$\left(\xi^{(0\sim m-1)}\right)^T P_m \rho(\xi^{(0\sim m-1)}) H^T(\xi^{(0\sim m-1)})\theta$$

$$\leq \|\left(\xi^{(0\sim m-1)}\right)^T P_m \rho(\xi^{(0\sim m-1)}) H^T(\xi^{(0\sim m-1)})\| \|\theta\| \tag{7.43}$$

$$\leq \eta\epsilon\left(\xi^{(0\sim m-1)}\right)^T P_m \rho(\xi^{(0\sim m-1)})\delta(\xi^{(0\sim m-1)}) + \eta\epsilon\varepsilon_2.$$

Consequently, (7.41) becomes

$$\dot{V} \leq -\frac{\mu}{2}\left(\xi^{(0\sim m-1)}\right)^T P\xi^{(0\sim m-1)} + \left(\xi^{(0\sim m-1)}\right)^T P_m\left(\rho(\xi^{(0\sim m-1)})\breve{\breve{u}}\right.$$

$$+ \eta\tilde{\epsilon}\rho(\xi^{(0\sim m-1)})(\zeta(\xi^{(0\sim m-1)}) + \delta(\xi^{(0\sim m-1)})) \tag{7.44}$$

$$\left. + \rho(\xi^{(0\sim m-1)})G(x^{(0\sim m-1)})u\right) - \underline{g}\tilde{v}\dot{\tilde{v}} - \eta\tilde{\epsilon}\dot{\tilde{\epsilon}} + (s + \eta\epsilon)\varepsilon_2.$$

It can be known from (7.34) that

$$\breve{d}_i(t) = (1 + \kappa_1(t)a)u_i(t) + \kappa_2(t)p_i, \tag{7.45}$$

where $\kappa_1(t)$ and $\kappa_2(t)$ are time-varying parameters satisfying $|\kappa_1(t)| \leq 1$ and $|\kappa_2(t)| \leq 1$. It means that

$$\breve{d}(t) = (1 + \kappa_1(t)a)u(t) + \kappa_2(t)p. \tag{7.46}$$

Therefore, we have

$$u = \frac{\breve{d}}{1 + \kappa_1 a} - \frac{\kappa_2 p}{1 + \kappa_1 a}. \tag{7.47}$$

Noting that $\left(\xi^{(0\sim m-1)}\right)^T P_m \rho(\xi^{(0\sim m-1)})G(\xi^{(0\sim m-1)})\breve{d} \leq 0$ and Lemma 2.8, then one has

$$\left(\xi^{(0\sim m-1)}\right)^T P_m \rho(\xi^{(0\sim m-1)})G(\xi^{(0\sim m-1)})u$$

$$= \frac{\left(\xi^{(0\sim m-1)}\right)^T P_m \rho(\xi^{(0\sim m-1)})G(\xi^{(0\sim m-1)})\breve{d}}{1 + \kappa_1 a}$$

$$- \frac{\kappa_2\left(\xi^{(0\sim m-1)}\right)^T P_m \rho(\xi^{(0\sim m-1)})G(\xi^{(0\sim m-1)})p}{1 + \kappa_1 a}$$

$$\leq \overline{g}|\frac{\left(\xi^{(0\sim m-1)}\right)^T P_m \rho(\xi^{(0\sim m-1)})p}{1 - a}|$$

$$+ \underline{g}\frac{\left(\xi^{(0\sim m-1)}\right)^T P_m \rho(\xi^{(0\sim m-1)})\breve{d}}{1 + a}$$

$$\leq \eta \epsilon (\xi^{(0 \sim m-1)})^T P_m \rho(\xi^{(0 \sim m-1)}) r_2(\xi^{(0 \sim m-1)})$$
$$+ \underline{g} \frac{\left(\xi^{(0 \sim m-1)}\right)^T P_m \rho(\xi^{(0 \sim m-1)}) \check{d}}{1 + a} + \varepsilon_1 \bar{g}. \tag{7.48}$$

From (7.37), it can be known that $\hat{\epsilon} \geq 0$ for all $t \geq 0$ when $\hat{\epsilon}(0) \geq 0$. Therefore,

$$\underline{g} \frac{\left(\xi^{(0 \sim m-1)}\right)^T P_m \rho(\xi^{(0 \sim m-1)}) \check{d}}{1 + a}$$
$$\leq -\underline{g} \hat{v}(\xi^{(0 \sim m-1)})^T P_m \rho(\xi^{(0 \sim m-1)}) \check{\bar{u}} + \varepsilon_1 \underline{g}$$
$$- \underline{g} \hat{\epsilon}(\xi^{(0 \sim m-1)})^T P_m \rho(\xi^{(0 \sim m-1)}) r_2 \tag{7.49}$$
$$\leq -\underline{g} \hat{v}(\xi^{(0 \sim m-1)})^T P_m \rho(\xi^{(0 \sim m-1)}) \check{\bar{u}} + \varepsilon_1 \underline{g}$$
$$- \eta \hat{\epsilon}(\xi^{(0 \sim m-1)})^T P_m \rho(\xi^{(0 \sim m-1)}) r_2.$$

Substituting (7.49) into (7.48), we obtain

$$\left(\xi^{(0 \sim m-1)}\right)^T P_m \rho(\xi^{(0 \sim m-1)}) G(\xi^{(0 \sim m-1)}) u$$
$$+ \underline{g} v(\xi^{(0 \sim m-1)})^T P_m \rho(\xi^{(0 \sim m-1)}) \check{\bar{u}}$$
$$\leq \underline{g} \tilde{v}(\xi^{(0 \sim m-1)})^T P_m \rho(\xi^{(0 \sim m-1)}) \check{\bar{u}} + \varepsilon_1 \underline{g} \tag{7.50}$$
$$+ \varepsilon_1 \bar{g} + \eta \tilde{\epsilon}(\xi^{(0 \sim m-1)})^T P_m \rho(\xi^{(0 \sim m-1)}) r_2.$$

By plugging (7.50) into (7.41), one has

$$\dot{V} \leq -\frac{\mu}{2} \left(\xi^{(0 \sim m-1)}\right)^T P \xi^{(0 \sim m-1)} + \varepsilon_1 \underline{g} + \varepsilon_1 \bar{g} + (s + \eta \epsilon) \epsilon_2$$
$$- \eta \tilde{\epsilon} \left(\dot{\hat{\epsilon}} - \left(\xi^{(0 \sim m-1)}\right)^T P_m \left(\rho(\xi^{(0 \sim m-1)})(\zeta(\xi^{(0 \sim m-1)})\right.\right.$$
$$+ \delta(\xi^{(0 \sim m-1)})) + \rho(\xi^{(0 \sim m-1)}) r_2 \Big) \Big) \tag{7.51}$$
$$- \underline{g} \tilde{v}(\dot{\hat{v}} - (\xi^{(0 \sim m-1)})^T P_m \rho(\xi^{(0 \sim m-1)}) \check{\bar{u}}.$$

It follows from (7.51) and (7.37) that

$$
\begin{aligned}
\dot{V} \leq & -\frac{\mu}{2}\left(\xi^{(0 \sim m-1)}\right)^{T} P \xi^{(0 \sim m-1)} - \frac{\sigma_1 \underline{g}}{2} \tilde{v}^2 \\
& -\frac{\sigma_2 \eta}{2} \tilde{\epsilon}^2 + \frac{\sigma_1 \underline{g}}{2} v^2 + \varepsilon_1 \underline{g} + \varepsilon_1 \bar{g} \\
& +\frac{\sigma_2 \eta}{2} \epsilon^2 + (s + \eta \epsilon)\varepsilon_2 \\
\leq & -cV + q.
\end{aligned}
\tag{7.52}
$$

where

$$
c = \min\{\mu, \sigma_1, \sigma_2\},
$$

$$
q = \frac{\sigma_1 \underline{g}}{2} v^2 + \varepsilon_1 \underline{g} + \varepsilon_1 \bar{g} + \frac{\sigma_2 \eta}{2} \epsilon^2 + (s + \eta \epsilon)\varepsilon_2.
$$

It is straightforward to obtain from (7.52) that

$$
V \leq e^{-ct}\left(V(0) - \frac{q}{c}\right) + \frac{q}{c}.
\tag{7.53}
$$

It means that

$$
V \to \frac{q}{c},
\tag{7.54}
$$

when $t \to +\infty$.

Then, $\xi^{(j)}$, $j = 0, \ldots, m-1$, $\hat{v}$, $\hat{\epsilon}$, $\varpi(\xi^{(0 \sim m-1)})$ are bounded. It can be known from (7.34)–(7.36), that $\check{d}$, u and $\check{u}$ are bounded. Together with (7.5) and (7.6), we can drive that $x^{(j)}$ are bounded. Therefore, all signals of the closed-loop system are bounded. Besides, it can be concluded from (7.2) that the boundedness of ξ_i implies $-\underline{\varphi}_i < x_i < \bar{\varphi}_i$.

Next, we will prove that the proposed ETM does not exhibit Zeno behavior. Suppose that there is a time interval $\underline{t} = [\underline{t}_1, \ldots, \underline{t}_r]^T \in \mathbb{R}^r$ satisfying

$$
\underline{t} \leq t_{k+1} - t_k
\tag{7.55}
$$

for $k \in \mathbb{N}_+$.

For $t \in [t_{i,k}, t_{i,k+1})$, the following equation holds

$$
\check{e}_i(t) = \check{d}_i(t) - u_i(t).
\tag{7.56}
$$

It implies that

$$\frac{d|\breve{e}_i|}{dt} = \text{sign}(\breve{e}_i)\dot{\breve{e}}_i \le |\dot{\breve{d}}_i|. \tag{7.57}$$

Because $\varpi(\xi^{(0\sim m-1)})$ and $H(\xi^{(0\sim m-1)})$ possess the first-order continuous partial derivatives, it is straightforward to conclude that $\dot{d}$ is continuous. Combined with the boundedness of all closed-loop signals, there must be a constant $\breve{\varsigma}_i \in \mathbb{R}_+$ such that

$$|\dot{\breve{d}}_i| \le \breve{\varsigma}_i. \tag{7.58}$$

From (7.35), it can be obtain that $\breve{e}_i(t_{i,k}) = 0$ and $\breve{e}_i(t_{i,k+1}) = a|u_i| + p_i$ hold. These derive that

$$\underline{t}_i \ge \frac{a|u_i| + p_i}{\breve{\varsigma}_i} \ge \frac{p_i}{\breve{\varsigma}_i}. \tag{7.59}$$

Therefore, the time interval $\underline{t}$ satisfies

$$\underline{t} \ge \breve{\omega}, \tag{7.60}$$

where $\breve{\omega} \in \mathbb{R}_+$. This means that the proposed ETM is Zeno-free.

The proof is complete.

Remark 7.7 Unlike in the previous chapter where adaptive laws need to be designed to estimate each unknown constant parameter involved in the controller design, this chapter proposed a simplified adaptive mechanism which is constructed by means of inequality scaling and norm estimation approach. This enables the control scheme to involve only two parameter adaptations during the system operation, significantly reducing the computational complexity.

To present the workflow of the proposed control algorithm more clearly, the following table is presented

7.4 Simulation Examples

The following RLC circuit is employed to verify the effectiveness of the designed controller (Fig. 7.2)

Based on the Kirchhoff's Law of Voltage and Current, the considered RLC circuit can be modeled as

$$LC\frac{d^2 u_c}{dt^2} + \Delta(u_c, \frac{du_c}{dt}) + RC\dot{x} + u_c = u_r, \tag{7.61}$$

Fig. 7.2 The RLC circuit

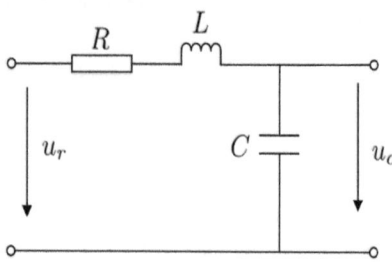

Algorithm of adaptive event-triggered control

Initialization:

(i) Check whether the HOFA system (7.1) complies with Assumptions 7.2–7.3.

(ii) Set the initial values: $x^{(0\sim m-1)}(0)$, $\hat{v}(0)$, $\hat{e}(0)$;

(iii) Determine the output constraints boundaries and check Assumption 7.1.

(iv) Based on Lemma 2.6, calculate the matrix P once given $A_0, \ldots, A_{m-1}$;

While $t < T$, where T denotes the lifespan of the HOFA system (7.1)

    Do   Calculate $d_i(t)$ and the measurement error $e_i(t)$

      For $i = \mathbb{N}_{1:r}$

      If The ETM (7.35) is satisfied

        (i) Set $t_{i,k+1} = t$, $u_i(t) = u_i(t_{i,k+1})$, and $e_i(t) = 0$;

        (ii) Send $u_i(t)$ to actuator.

      Else $u_i(t) = u_i(t_{i,k})$, and $e_i(t) = d_i(t) - u_i(t_{i,k})$.

End All

In this simulation, we take $L = 0.5(H)$, $C = 1(F)$, $R = 0.5(\Omega)$ and $\Delta(x, \dot{x}) = -(0.25 \sin(x^2) + 0.5\dot{x}^2)$.

By taking $x = u_c$, $\dot{x} = \frac{du_c}{dt}$, $\ddot{x} = \frac{d^2 u_c}{dt^2}$ and $u = u_r$, we can rewritten (7.61) as

$$\ddot{x} = G(x^{(0\sim 1)})u + f(x^{(0\sim 1)}) + \Delta f(x^{(0\sim 1)}) + H^T(x^{(0\sim 1)})\theta,$$

where $G(x^{(0\sim 1)}) = 2$, $f(x^{(0\sim 1)}) = -2\dot{x}$, $\Delta f(x^{(0\sim 1)}) = 0.5 \sin(x^2) + \dot{x}^2$, $H(x^{(0\sim 1)}) = 2x$, $\theta = 1$. The initial values are taken as $x(0) = 2$, $\dot{x}(0) = -3$.

Based on Assumption 2, we choose $\varpi(x^{(0\sim 1)}) = 0.5x^2 + \dot{x}^2$. From Lemma 2.6, we select $A^{0\sim 1} = [5, 10]$, $\Gamma = I_{2\times 2}$, $P_2 = [1, 2]^T$.

For the adaptive event-triggered controller (7.8), the initial values of adaptive controller are taken as $\hat{\theta}(0) = 0.5$, $\hat{s}(0) = -0.5$, $\hat{v}(0) = 0.6$, $\hat{b}(0) = 0.8$. The other parameters are $\sigma_1 = 5$, $\sigma_2 = 5$, $\sigma_3 = 1$, $\varepsilon_1 = 0.0001$, $\varepsilon_2 = 0.0001$, $a = 0.2$, $p = 0.5$. Figures 7.3, 7.4, 7.5, and 7.6 show the simulation results. Figure 7.3a shows that the system state x remains within the predefined range $(-\varphi, \overline{\varphi})$. Figure 7.3b is the resopose of $\dot{x}$ which is also bounded. The trajectories of the virtual control input d and actual control input u are displayed in Fig. 7.4a which confirms the stability of the control scheme. Figure 7.4b depicts the event-triggered time. From it, we can known that there is no Zeno phenomenon. Figures 7.5 and 7.6 illustrates the

Fig. 7.3 The responses of x
and $\dot{x}$. (**a**) The response of x.
(**b**) The response of $\dot{x}$

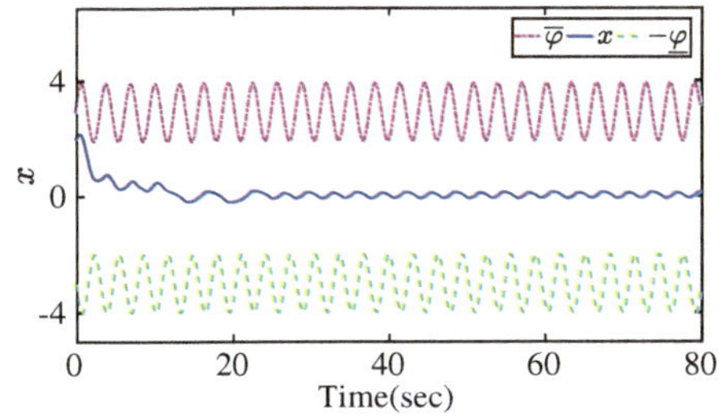

(a) The response of x

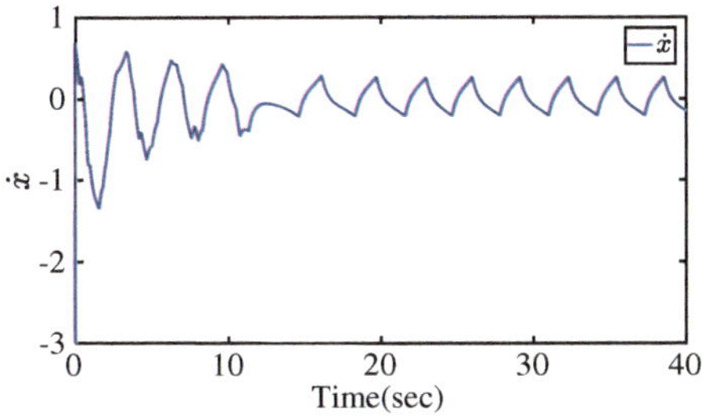

(b) The response of $\dot{x}$

Fig. 7.4 The responses of u,
d and t_k. (**a**) The responses of
u, d. (**b**) The response of t_k

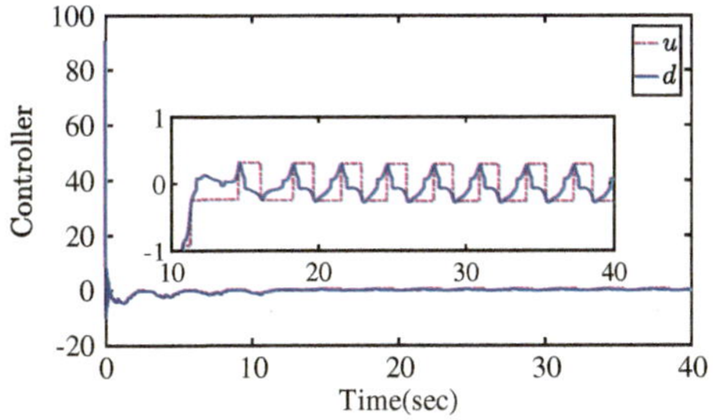

(a) The responses of u, d

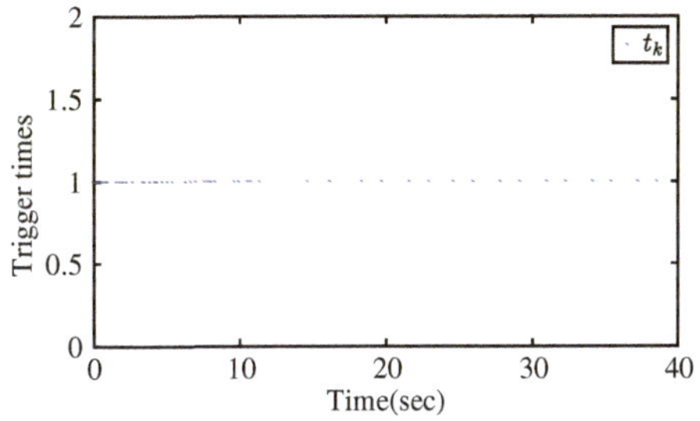

(b) The response of t_k

convergence performance of the designed adaptive laws $\hat{\theta}$, $\hat{s}$, $\hat{v}$ and $\hat{b}$, respectively.
Obviously, the proposed algorithm is effective.

For the adaptive event-triggered controller (7.34), the initial values of adaptive
controller are taken as $\hat{v}(0) = 0.6$ and $\hat{\epsilon}(0) = 0.5$. The other parameters are $\sigma_1 =$

Fig. 7.5 The responses of $\hat{\theta}$, and $\hat{s}$. (**a**) The response of $\hat{\theta}$. (**b**) The response of $\hat{s}$

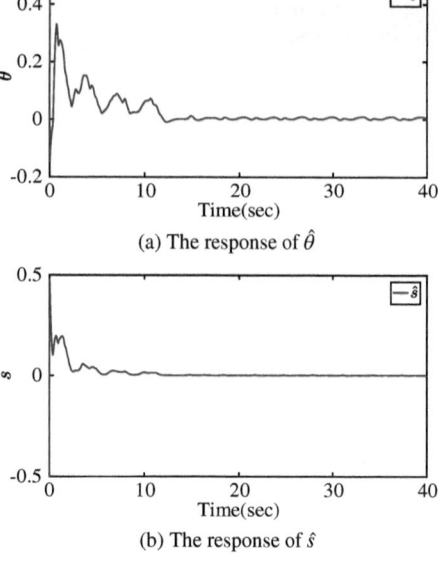

(a) The response of $\hat{\theta}$

(b) The response of $\hat{s}$

Fig. 7.6 The responses of $\hat{v}$, and $\hat{b}$. (**a**) The response of $\hat{v}$. (**b**) The response of $\hat{b}$

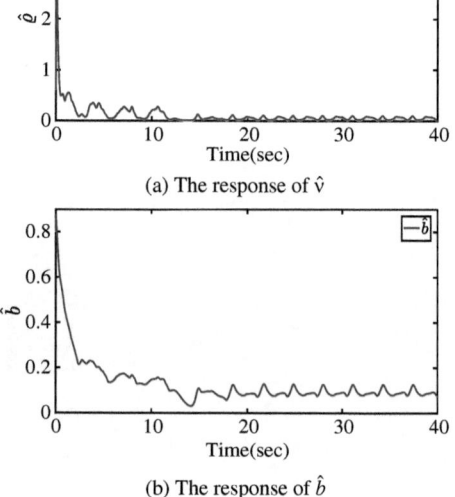

(a) The response of $\hat{v}$

(b) The response of $\hat{b}$

$5, \sigma_2 = 5, \sigma_3 = 1, \varepsilon_1 = 0.0001, \varepsilon_2 = 0.0001, a = 0.2, p = 0.5$. The simulation results are shown in Figs. 7.7, 7.8, and 7.9. As shown in Fig. 7.7a, the system state x satisfies the predefined constraints. From Fig. 7.7b which is the response of $\dot{x}$, we can see that $\dot{x}$ is also bounded. Figure 7.8a displays the trajectories of the virtual control input d and actual control input u. From it, we can confirm the stability

Fig. 7.7 The responses of x
and $\dot{x}$. (**a**) The response of x.
(**b**) The response of $\dot{x}$

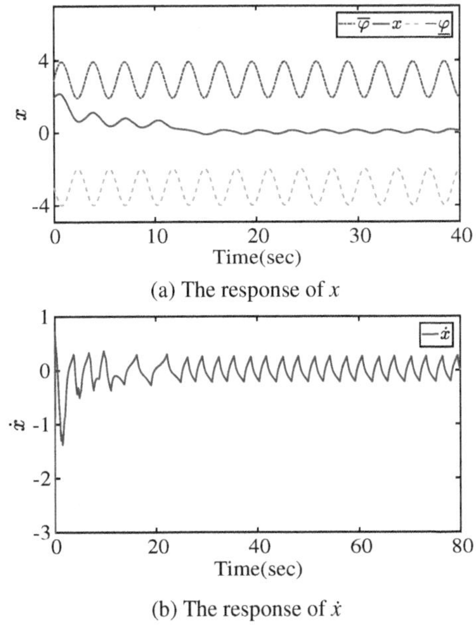

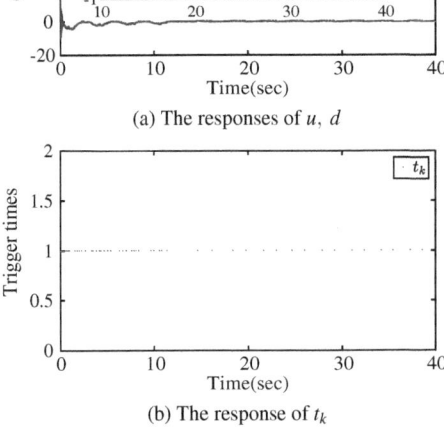

(a) The response of x

(b) The response of $\dot{x}$

Fig. 7.8 The responses of u,
d and t_k. (**a**) The responses of
u, d. (**b**) The response of t_k

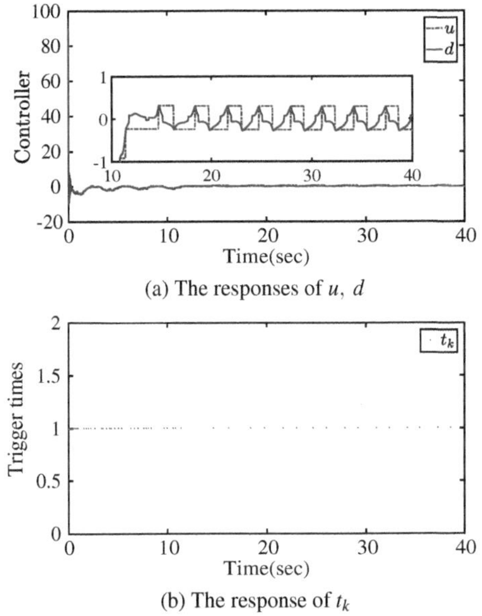

(a) The responses of u, d

(b) The response of t_k

of the control scheme. It is seen from Fig. 7.8b that there is no Zeno phenomenon. Figure 7.9 illustrates the convergence performance of the designed adaptive laws $\hat{\theta}$ and $\hat{\epsilon}$, respectively. It is not difficult to see that the controller is effective. Besides, it can also be observed from the comparison of Figs. 7.3, 7.4, 7.5, 7.6, 7.7, 7.8, and 7.9 that even when the parameters $A^{0\sim1}$, Γ, P_2, ε_1, ε_2, a and p are the same, the

Fig. 7.9 The responses of $\hat{v}$, and $\hat{\epsilon}$. (**a**) The response of $\hat{v}$. (**b**) The response of $\hat{\epsilon}$

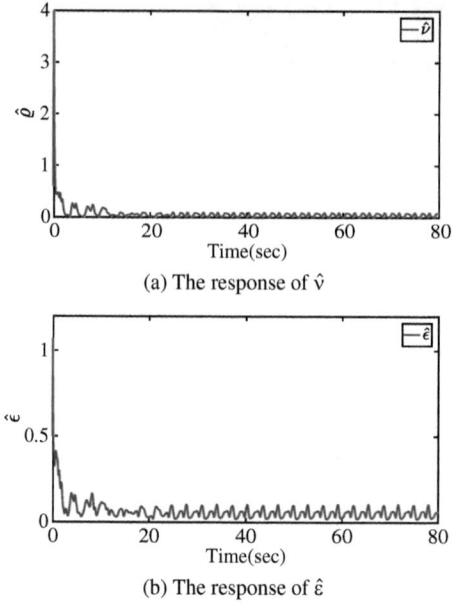

(a) The response of $\hat{v}$

(b) The response of $\hat{\epsilon}$

system responses are also different. Although the computational load can be reduced by using a lower adaptive rate in the controller, it may result in a longer stabilization time and a larger maximum control input.

7.5   Notes and References

This chapter studies the adaptive event-triggered control problem for a class of HOFA systems with output constraints. To address the problem of unknown control matrix functions, an adaptive technique is adopted. To save energy consumption in signal transmission, the controller and its event-triggered mechanism are designed. By using Lyapunov stability theory, it is rigorously proved that the proposed controller can ensure that the output of the system can be constrainted in the asymmetric time-varying boundaries and all signals of the closed-loop system are bounded. Finally, based on numerical simulations of an RLC circuit, the validity of the theoretical results is verified.

Chapter 8
Multi-Variable Constrained Control for High-Order Fully Actuated Nonlinear Systems

Expanding on the constrained control techniques from the previous chapter, this chapter addresses multi-variable HOFA nonlinear systems with uncertainties. These systems often involve multiple interacting subsystems, requiring coordinated control while respecting constraints. We introduce a novel control strategy that integrates command filtered control with HOFA system theory, managing multi-variable interactions and addressing uncertainties using radial basis function neural networks.

8.1 Overview

In the domain of nonlinear control systems, strict-feedback architectures (SFNS) serve as fundamental models for characterizing real-world nonlinear phenomena. Significant advancements have been made in SFNS research via backstepping techniques (Jia et al. 2022; Yu et al. 2020; Tong et al. 2020; Yang et al. 2018), yet existing studies predominantly focus on first-order system formulations. Practical systems governed by Newtonian or Lagrangian dynamics inherently exhibit HOFA structures. Traditional control paradigms often decompose these systems into cascaded first-order subsystems, introducing computational burdens. To overcome this limitation, the HOFA framework (Duan 2021a,b,e) enables direct controller synthesis for native high-order models or transformed HOFA representations. This methodology facilitates nonlinearity cancellation to establish linear closed-loop dynamics (Zhang et al. 2023e). Subsequent innovations integrate backstepping with HOFA for robust/adaptive control of uncertain HOSF systems (Duan 2021c,d), albeit requiring structured uncertainty assumptions. For unstructured uncertainties in HOSF systems, conventional adaptive/robust strategies face limitations, prompting explorations of fuzzy logic (FLS) and neural networks (NN) (Wang et al. 2022d; Ma et al. 2021; Sun et al. 2020a). While Cui et al. (2023a) combines HOFA with

© The Author(s) 2026
C. Hua et al., *Adaptive Constrained Control for High-Order Fully Actuated Nonlinear Systems*, https://doi.org/10.1007/978-981-95-0962-1_8

FLS for fault-tolerant control, persistent challenges remain in virtual controller differentiation. Recent work Liu et al. (2023b,c) introduces command-filtered backstepping within HOFA frameworks to mitigate differentiation complexity, though multivariable constraint handling in HOSF systems warrants further investigation.

State-constrained control research bifurcates into barrier Lyapunov function (BLF) and non-barrier BLF (NBLF) methodologies (Li et al. 2017c; Liang et al. 2020; Liu et al. 2018; Li et al. 2022c, 2024b). BLF variants (logarithmic/integral/tangent) enforce state constraints (Liu & Tong 2016; Li et al. 2017a; Ouyang & Lin 2021; Zhang et al. 2023f), while NBLF approaches (Zhao & Song 2019; Zhao et al. 2019b; Guo et al. 2023) employ nonlinear transformations to relax feasibility conditions. Both paradigms necessitate constrained initial states–a requirement often impractical for unknown initial conditions. Deferred constraint strategies (Song & Zhou 2018; Sun et al. 2021; Shao & Ye 2022) and singular transformation functions (Zhao et al 2022a; Cao et al. 2022b; Zhang et al. 2022b) address this limitation, with practical prescribed-time control (Cao et al. 2022b) enabling unconstrained initialization through predefined convergence time/accuracy. Recent extensions (Guo et al. 2023; Guo & Hu 2023b; Luo et al. 2024) enhance PPTC for high-order systems, though predominantly targeting first-order subsystem formulations.

For second-order SFNSs, Cao et al. (2022a) investigates PPTC in Euler-Lagrange systems via first-order decomposition–a process increasing computational complexity. HOFA-based studies (Li et al. 2022e; Xiao et al. 2024; Dong et al. 2023; Wang & Duan 2024) simplify constrained control design for HOSF systems, yet Multi-variable constraint management and complexity reduction in fully actuated nonlinear systems remain open challenges.

Motivated by these observations, this chapter aims to explore the HOFA system approach design of multi-variable constrained control for uncertain HOSF fully actuated nonlinear systems. The new features of this study are outlined as follows:

1. Unlike prior studies limited to output constraints in HOSF fully actuated systems (Li et al. 2022e; Xiao et al. 2024) or full-state constraints for single HOFA systems (Dong et al. 2023; Wang & Duan 2024), our strategy achieves constrained control for internal state variables (beyond outputs) in HOSF systems. Additionally, it removes dependence on initial state knowledge and constrained initialization requirements.
2. We pioneer the fusion of command filtering, RBF neural networks, and PPTC within the HOFA framework. This synthesis resolves persistent virtual controller differentiation challenges and manages unstructured uncertainties in constrained HOSF systems with optimized performance–advances unaddressed in Cui et al. (2023a) and Liu et al. (2023b).
3. Compared to existing state-constrained approaches for high-order SFNSs (Li et al. 2017a; Sun et al. 2021; Cao et al. 2022a), our HOFA-based design delivers superior efficacy and operational simplicity without proportional growth in computational overhead. This streamlined methodology significantly reduces implementation complexity–a critical advantage for real-world deployment.

8.2 System Formulation and Preliminaries

8.2.1 System Formulation

Consider an uncertain HOSF fully actuated nonlinear system as

$$
\begin{cases}
q_i^{(m_i)} = p_i(q_j^{(0\sim m_j-1)}|_{j=1\sim i})q_{i+1} + l_i(q_j^{(0\sim m_j-1)}|_{j=1\sim i}), \\
q_n^{(m_n)} = p_n(q_j^{(0\sim m_j-1)}|_{j=1\sim n})u + l_n(q_j^{(0\sim m_j-1)}|_{j=1\sim n}),
\end{cases}
\tag{8.1}
$$

where $1 \le i \le n - 1$, $q_i \in \mathbb{R}$, $u \in \mathbb{R}$ represent system state variables and control input. $p_i(q_j^{(0\sim m_j-1)}|_{j=1\sim i})$, $i = 1, \ldots, n$ are known smooth nonlinearities with $p_i(q_j^{(0\sim m_j-1)}|_{j=1\sim i}) \ne 0$, and $l_i(q_j^{(0\sim m_j-1)}|_{j=1\sim i})$, $i = 1, \ldots, n$ are unknown nonlinearities with $l_i(0) = 0$. m_i, $i = 1, \ldots, n$ are positive integers.

The control objective is to design a controller u to ensure that the system state variables q_i converge to the specified invariant region within a predetermined time, regardless of the initial conditions, while also guaranteeing boundedness of all signals in the closed-loop system. In the following, we will introduce some key lemmas.

Based on Lemmas 2.5–2.6, we define a set of matrices $A_i^{0\sim m_i-1} \in \mathbb{R}^{1\times m_i}$, $i = 1, \ldots, n$ to make $\Phi(A_i^{0\sim m_i-1})$, $i = 1, \ldots, n$ stable and

$$
P_{iL}(A_i^{0\sim m_i-1}) = P_i(A_i^{0\sim m_i-1}) \begin{bmatrix} 0_{(m_i-1)\times 1} \\ I \end{bmatrix},
$$

which will be utilized in subsequent analysis.

Lemma 8.1 (Liu et al. (2023b)) *The command filter is defined as*

$$
\begin{aligned}
\dot{\xi}_1 &= g\xi_2, \\
\dot{\xi}_2 &= -2\varpi g\xi_2 - g(\xi_1 - \alpha_1).
\end{aligned}
$$

If the input signal α_1 satisfies $|\dot{\alpha}_1| \le o_1$ and $|\ddot{\alpha}_1| \le o_2$ for all $t > 0$, where o_1, o_2 are positive constants and $\xi_1(0) = \alpha_1(0)$, $\xi_2(0) = 0$, then for any $\kappa > 0$, there exist $0 < \varpi \le 1$, $g > 0$, such that $|\xi_2 - \alpha_1| \le \kappa$, $|\dot{\xi}_1|$, $|\ddot{\xi}_1|$ and $|\dddot{\xi}_1|$ are bounded.

Proposition 8.1 *For any arbitrary $F \in \mathbb{R}^{m_i\times m_i}$, all the matrix $A_i^{0\sim m_i-1}$ and the nonsingular matrix $V \in \mathbb{R}^{m_i\times m_i}$ satisfying $\Phi(A_i^{0\sim m_i-1}) = VFV^{-1}$ are determined by*

$$
A_i^{0\sim m_i-1} = -ZF^{m_i}V^{-1}(Z, F), \quad V(Z, F) = \begin{bmatrix} Z & ZF & \cdots & ZF^{m_i-1} \end{bmatrix}^T,
$$

where $Z \in \mathbb{R}^{1\times m_i}$ is a designed parameter matrix satisfying $\det V(Z, F) \ne 0$.

8.2.2   Adaptive RBF Neural Network

For any continuous nonlinear function $F(\Xi)$, the RBF neural network can approximate it over a compact set $\Omega_\Xi \subset R^q$ with an arbitrary accuracy $\bar{\varepsilon} > 0$, which is described as

$$F(\Xi) = \theta^T W(\Xi) + \varepsilon, \ |\varepsilon| \le \bar{\varepsilon},$$

where $\Xi \in \Omega_\Xi \subset R^q$ is the input vector with q as the network input dimension. ε represents the approximation error. The basis function vector $W(\Xi) = [W_1(\Xi), \cdots, W_l(\Xi)]^T$ with $l > 1$ neural network nodes is defined as

$$W_i(\Xi) = \exp\left(\frac{-\|(\Xi - \varsigma_i)\|^2}{\varrho^2} \right),$$

where $\varsigma_i = [\varsigma_{i1}, \cdots, \varsigma_{iq}]^T$ and ϱ are the center and width of the Gaussian function, respectively. The ideal weight vector $\theta = [\theta_1, \cdots, \theta_l]^T \in R^l$ minimizes ε for all $\Xi \in \Omega_\Xi$ given by

$$\theta = \arg\min_{\hat{\theta} \in R^l} \left\{ \sup_{\Xi \in \Omega_\Xi} \left| F(\Xi) - \hat{\theta}^T W(\Xi) \right| \right\},$$

where $\hat{\theta}$ is the weight value vector.

Remark 8.1 The nonlinear system (8.1) is classified as high-order when $m_i > 1$ and traditional first-order when $m_i = 1$ ($i = 1, \ldots, n$). This formulation broadly encapsulates practical engineering systems governed by Newtonian dynamics, Lagrangian mechanics, and electromechanical principles. A key requirement for applying the HOFA framework is the non-singularity of the control gain $p_i(q_j^{(0 \sim m_j - 1)}|_{j=1 \sim i})$, a condition widely acknowledged in prior research. Each subsystem's fully actuated nature ($p_i \neq 0$) eliminates the need for order reduction, enabling direct controller synthesis that yields linear closed-loop dynamics. The main focus of this study is to employ the HOFA system approach to ensure that all system state variables q_i ($i = 1, \ldots, n$) adhere to the desired constraint restrictions. Proposition 8.1 details the linear subsystem design methodology for closed-loop implementation, extending principles from Duan (2021c,d).

8.3   Controller Design Incorporating Command Filtering Techniques

8.3.1   Constraint Function Design and Coordinate Transformation

As outlined in Cao et al. (2022b), the PPTC strategy is introduced to achieve the following objective

$$- \rho_i(t) < q_i(t) < \rho_i(t), \tag{8.2}$$

where the constraint function is designed as

$$\rho_i(t) = \begin{cases} \left(\frac{1}{t} - \frac{1}{T_f} \right)^{2p} + \rho_{iT_f}, & t \in (0, T_f], \\ \rho_{iT_f}, & t \in (T_f, \infty), \end{cases}$$

with positive designed parameters ρ_{iT_f}, T_f, and positive integer p satisfying $2p > n + 1$.

Then, construct a compound function $h_i(\gamma_i)$ as

$$h_i(\gamma_i) = \begin{cases} 1 - \left(\frac{\gamma_i}{a_i} - 1 \right)^{2p}, & 0 < \gamma_i \leq a_i, \\ 1, & \gamma_i > a_i, \end{cases} \tag{8.3}$$

where $a_i > c_i \rho_{iT_f}^2$, $\gamma_i = c_i(\rho_i - q_i)(\rho_i + q_i)$ with $c_i > 0$. With this construction, we define a transformation given by

$$\kappa_{i1} = \frac{q_i(t)}{h_i(\gamma_i)}, \tag{8.4}$$

and differentiating it yields

$$\dot{\kappa}_{i1} = r_{i1}\dot{q}_i + v_{i1}, \tag{8.5}$$

where

$$r_{i1} = \begin{cases} \frac{1}{h_i} - \frac{4c_i p}{a_i h_i^2} \left(\frac{\gamma_i}{a_i} - 1 \right)^{2p-1} q_i^2, & 0 < \gamma_i \leq a_i, \\ 1, & \gamma_i > a_i, \end{cases}$$

$$v_{i1} = \begin{cases} \frac{4c_i p}{a_i h_i^2} \left(\frac{\gamma_i}{a_i} - 1 \right)^{2p-1} \rho_i \dot{\rho}_i q_i, & 0 < \gamma_i \leq a_i, \\ 0, & \gamma_i > a_i. \end{cases}$$

Define $\kappa_{ij} = q_i^{(j-1)}$, $j = 2, \ldots, m_i$. It follows that

$$\begin{cases} \dot{\kappa}_{i1} = g_i^* \kappa_{i2} + \varphi_i^*, \\ \dot{\kappa}_{ij} = \kappa_{i,j+1}, \ j = 2, \ldots, m_i - 1, \\ \dot{\kappa}_{im_i} = p_i(\bar{\kappa}_{im_i}) q_{i+1} + l_i(\bar{\kappa}_{im_i}), \end{cases} \tag{8.6}$$

where $\bar{\kappa}_{im_i} = [\kappa_{i1}, \kappa_{i2}, \ldots, \kappa_{im_i}]^T$, $g_i^*(q_i, t) = r_{i1}$, and $\varphi_i^*(q_i, t) = v_{i1}$.
Construct the following coordinate transformation

$$\begin{aligned} x_i &= \kappa_{i1}, \\ \kappa_{i2} &= \left(g_i^*\right)^{-1} (\dot{x}_i - \varphi_i^*), \\ \kappa_{ij} &= \left(g_i^*\right)^{-1} (x_i^{(j-1)} - \phi_{i,j-1}^*), \ j = 2, \ldots, m_i - 1, \end{aligned} \tag{8.7}$$

where

$$\begin{aligned} \phi_{i,j-1}^*(x_i^{(0 \sim j-2)}, t) &= \dot{g}_i^* \kappa_{i,j-1} + \dot{\phi}_{i,j-2}^* \\ &= \dot{g}_i^* \kappa_{i,j-1} + (\dot{g}_i^* \kappa_{i,j-2})^{(1)} + (\dot{g}_i^* \kappa_{i,j-3})^{(2)} \\ &\quad + \cdots + (\dot{g}_i^* \kappa_{i,2})^{(j-3)} + (\varphi_i^*)^{(j-2)}. \end{aligned}$$

Then, we can obtain

$$x_i^{(2)} = \dot{g}_i^* \kappa_{i2} + g_i^* \dot{\kappa}_{i2} + \dot{\varphi}_i^* = g_i^* \kappa_{i3} + \phi_{i2}^*, \tag{8.8}$$

where $\phi_{i2}^*(x_i^{(0 \sim 1)}, t) = \dot{g}_i^* \kappa_{i2} + \dot{\varphi}_i^*$.
From (8.7), it is easy to obtain $\kappa_{i3} = \left(g_i^*\right)^{-1} (x_i^{(2)} - \phi_{i,2}^*)$. By taking the derivative of $x_i^{(2)}$ with respect to time, we get

$$x_i^{(3)} = \dot{g}_i^* \kappa_{i3} + g_i^* \dot{\kappa}_{i3} + \dot{\phi}_{i2}^* = g_i^* \kappa_{i4} + \phi_{i3}^*, \tag{8.9}$$

where $\phi_{i3}^*(x_i^{(0 \sim 2)}, t) = \dot{g}_i^* \kappa_{i3} + \dot{\phi}_{i2}^* = \dot{g}_i^* \kappa_{i,3} + (\dot{g}_i^* \kappa_{i,2})^{(1)} + (\varphi_i^*)^{(2)}$.
Similarly, we can recursively obtain $\kappa_{ij} = \left(g_i^*\right)^{-1} (x_i^{(j-1)} - \phi_{i,j-1}^*)$. The derivative of $x_i^{(j-1)}$ is

$$x_i^{(j)} = \dot{g}_i^* \kappa_{ij} + g_i^* \dot{\kappa}_{ij} + \dot{\phi}_{i,j-1}^* = g_i^* \kappa_{i,j+1} + \phi_{ij}^*, \ j = 4, \ldots, m_i - 1 \tag{8.10}$$

where

$$\begin{aligned} \phi_{ij}^*(x_i^{(0 \sim j-1)}, t) &= \dot{g}_i^* \kappa_{ij} + \dot{\phi}_{i,j-1}^* \\ &= \dot{g}_i^* \kappa_{i,j} + (\dot{g}_i^* \kappa_{i,j-1})^{(1)} + (\dot{g}_i^* \kappa_{i,j-2})^{(2)} + \cdots + (\dot{g}_i^* \kappa_{i,2})^{(j-2)} + (\varphi_i^*)^{(j-1)}. \end{aligned}$$

When $j = m_i$, we can obtain

$$
\begin{aligned}
x_i^{(m_i)} &= \dot{g}_i^* \kappa_{im_i} + g_i^* \dot{\kappa}_{im_i} + \dot{\phi}_{i,m_i-1}^* \\
&= \dot{g}_i^* \left(p_i(\bar{\kappa}_{im_i}) q_{i+1} + l_i(\bar{\kappa}_{im_i}) \right) + g_i^* \dot{\kappa}_{im_i} + \dot{\phi}_{i,m_i-1}^* \\
&= g_i h_{i+1}(\gamma_{i+1}(t)) x_{i+1} + \phi_i + f_i,
\end{aligned}
\tag{8.11}
$$

where

$$
g_i(x_j^{(0\sim m_j-1)}|_{j=1\sim i}, t) = \dot{g}_i^* p_i(\bar{\kappa}_{im_i}),
$$

$$
\phi_i(x_j^{(0\sim m_j-1)}|_{j=1\sim i}, t) = \dot{g}_i^* \kappa_{im_i} + \dot{\phi}_{i,m_i-1}^*,
$$

$$
f_i(x_j^{(0\sim m_j-1)}|_{j=1\sim i}, t) = \dot{g}_i^* l_i(\bar{\kappa}_{im_i}).
$$

With the above analysis, we can get the transformed system (8.1) in the form of

$$
\begin{cases}
x_i^{(m_i)} = g_i(x_j^{(0\sim m_j-1)}|_{j=1\sim i}, t) h_{i+1} x_{i+1} + \phi_i(x_j^{(0\sim m_j-1)}|_{j=1\sim i}, t) \\
\qquad + f_i(x_j^{(0\sim m_j-1)}|_{j=1\sim i}, t), \quad i = 1, \dots, n-1, \\
x_n^{(m_n)} = g_n(x_j^{(0\sim m_j-1)}|_{j=1\sim n}, t) u + \phi_n(x_j^{(0\sim m_j-1)}|_{j=1\sim n}, t) \\
\qquad + f_n(x_j^{(0\sim m_j-1)}|_{j=1\sim n}, t).
\end{cases}
\tag{8.12}
$$

Remark 8.2 The constraint transformation (8.4) with compound function $h_i(\gamma_i)$ serves as a critical mechanism for PPTC implementation without initial state constraints. Given $\rho_i(0) = \infty$ in (8.2), $\gamma_i(0) = \infty$ holds inherently. Owing to the structural definition of $h_i(\gamma_i)$ in (8.3), we obtain $h_i(\gamma_i(0)) = 1$ since $\rho_i(0) > a_i$, ensuring $\kappa_{i1} = q_i(t)$ at initialization–thereby fulfilling the design objective. Conversely, traditional transformation functions like $\tanh^{-1}(q_i/\rho_i) = \frac{1}{2}\ln[(\rho_i + q_i)/(\rho_i - q_i)]$ exhibit critical limitations: Their derivatives $(\frac{1}{2}\ln[(\rho_i + q_i)/(\rho_i - q_i)])' = \rho_i(\dot{q}_i - q_i\dot{\rho}_i/\rho_i)/(\rho_i^2 - q_i^2)$ cause vanished control gains initially $((\rho_i(0))/(\rho_i^2(0) - q_i^2(0)) = 0)$, resulting in initial system uncontrollability.

Remark 8.3 For state-constrained control of HOSF system (8.1), the PPTC method (Cao et al. 2022b) combined with transformations (8.4) and (8.7) achieves constraint-to-unconstraint conversion. This guarantees predefined convergence time/accuracy while maintaining the original system structure–a crucial requirement for subsequent HOFA controller design. The principal challenge resides in preserving the system's structural integrity during transformation. A key merit of transformed system (8.12) lies in its structural similarity to (8.1), with non-zero control gain $g_i(x_j^{(0\sim m_j-1)}|_{j=1\sim i}, t) h_{i+1}$ $(0 < h_{i+1} \leq 1)$ retaining full actuation properties. This structural preservation significantly simplifies the control synthesis process.

8.3.2 Command Filtering Construction

To facilitate control design, the following transformation of coordination is given

$$
\begin{aligned}
\zeta_1^{(0\sim m_1)} &= x_1^{(0\sim m_1)}, \\
\zeta_i^{(0\sim m_i)} &= x_i^{(0\sim m_i)} - \alpha_{if}^{(0\sim m_i)}, \, i = 2, \ldots, n
\end{aligned}
\tag{8.13}
$$

where α_{if} is the output of a series of novel command filters

$$
\begin{aligned}
\dot{\xi}_{ij} &= \xi_{i,j+1}, j = 1, \ldots, m_i - 1, \\
\dot{\xi}_{im_i} &= -\omega_{im_i}\xi_{im_i} - \cdots - \omega_{i2}\xi_{i2} - \omega_{i1}(\xi_{i1} - (\tfrac{\alpha_{i-1}}{h_i})),
\end{aligned}
\tag{8.14}
$$

in which α_{i-1} is the virtual input, $\omega_{ij} > 0$, $j = 1, \ldots, m_i$ are chosen to make $s^{(m_i)} + \omega_{im_i}s^{(m_i-1)} + \cdots + \omega_{i1} = 0$ stable. Additionally, we define $\alpha_{if}^{(j-1)} = \xi_{ij}$, $j = 1, \ldots, m_i$ and $\alpha_{if}^{(m_i)} = \dot{\xi}_{im_i}$ as each command filter's outputs.

The dynamic errors ψ_{ij} are defined as

$$
\psi_{ij} = \alpha_{if}^{(j-1)} - (\tfrac{\alpha_{i-1}}{h_i})^{(j-1)}, \, j = 1, \ldots, m_i,
\tag{8.15}
$$

then we get

$$
\begin{aligned}
\dot{\psi}_{ij} &= \psi_{i,j+1}, j = 1, \ldots, m_i - 1, \\
\dot{\psi}_{im_i} &= -\omega_{im_i}\psi_{im_i} - \omega_{i,m_i-1}\psi_{i,m_i-1} - \cdots - \omega_{i1}\psi_{i1} \\
&\quad - ((\tfrac{\alpha_{i-1}}{h_i})^{(m_i)} + \omega_{im_i}(\tfrac{\alpha_{i-1}}{h_i})^{(m_i-1)} \\
&\quad + \cdots + \omega_{i2}(\tfrac{\alpha_{i-1}}{h_i})^{(1)}), i = 2, \ldots, n,
\end{aligned}
\tag{8.16}
$$

and we can rewrite dynamic system (8.16) as

$$
\dot{\psi}_i = A_{\psi_i}\psi_i - \Delta_i,
\tag{8.17}
$$

where

$$
\psi_i = \begin{bmatrix} \psi_{i1} \\ \vdots \\ \psi_{im_i} \end{bmatrix}, \, A_{\psi_i} = \begin{bmatrix} 0 & 1 & 0 & 0 \\ \vdots & 0 & \ddots & 0 \\ 0 & 0 & 0 & 1 \\ -\omega_{i1} & -\omega_{i2} & \cdots & -\omega_{im_i} \end{bmatrix},
$$

$$
\Delta_i = \begin{bmatrix} 0_{(m_i-1)\times 1} \\ \left((\tfrac{\alpha_{i-1}}{h_i})^{(m_i)} + \omega_{im_i}(\tfrac{\alpha_{i-1}}{h_i})^{(m_i-1)} + \cdots + \omega_{i2}(\tfrac{\alpha_{i-1}}{h_i})^{(1)}\right) \end{bmatrix}.
$$

Given that A_{ψ_i} is Hurwitz, in accordance with Lemma 2.6, we can derive

$$A_{\psi_i}^T P_{\psi_i} + P_{\psi_i} A_{\psi_i} < -\mu_{\psi_i} P_{\psi_i}, \tag{8.18}$$

where $\mu_{\psi_i} > 0$ and $P_{\psi_i} \in \mathbb{R}^{n \times n}$ represent a positive constant and a positive definite matrix, respectively.

8.3.3 Backstepping Controller Design

In the following, we will apply the backstepping method to design controller. It contains n steps.

Step 1: Consider the following equations

$$\begin{aligned} \zeta_1^{(0\sim m_1)} &= x_1^{(0\sim m_1)}, \\ \zeta_2^{(0\sim m_2)} &= x_2^{(0\sim m_2)} - \alpha_{2,f}^{(0\sim m_2)}. \end{aligned} \tag{8.19}$$

Then it follows that

$$\begin{aligned} \zeta_1^{(m_1)} &= f_1(x_1^{(0\sim m_1-1)}, t) + \phi_1(x_1^{(0\sim m_1-1)}, t) + g_1(x_1^{(0\sim m_1-1)}, t)h_2(\zeta_2 + \alpha_{2f}) \\ &= F_1(\zeta_1^{(0\sim m_1-1)}, t) + \Psi_1(\zeta_1^{(0\sim m_1-1)}, t) + L_1(\zeta_1^{(0\sim m_1-1)}, t)h_2\zeta_2 \\ &\quad + L_1(\zeta_1^{(0\sim m_1-1)}, t)h_2\psi_{21} + L_1(\zeta_1^{(0\sim m_1-1)}, t)\alpha_1, \end{aligned} \tag{8.20}$$

where

$$\begin{aligned} F_1(\zeta_1^{(0\sim m_1-1)}, t) &= f_1(x_1^{(0\sim m_1-1)}, t), \\ \Psi_1(\zeta_1^{(0\sim m_1-1)}, t) &= \phi_1(x_1^{(0\sim m_1-1)}, t), \\ L_1(\zeta_1^{(0\sim m_1-1)}, t) &= g_1(x_1^{(0\sim m_1-1)}, t). \end{aligned}$$

According to the neural approximation, $F_1(\zeta_1^{(0\sim m_1-1)}, t)$ can be described by

$$F_1(\zeta_1^{(0\sim m_1-1)}, t) = \theta_1^T W_1(\zeta_1^{(0\sim m_1-1)}) + \varepsilon_1, \tag{8.21}$$

where ε_1 is the approximation error and satisfies $\varepsilon_1 < \bar{\varepsilon}_1$ with $\bar{\varepsilon}_1$ being a positive constant. Construct virtual controller α_1 and adaptive law $\hat{\theta}_1$ as

$$\begin{aligned} \alpha_1 &= -L_1^{-1}(\zeta_1^{(0\sim m_1-1)}, t)(A_1^{0\sim m_1-1}\zeta_1^{(0\sim m_1-1)} + \alpha_{11} + \alpha_{12}), \\ \alpha_{11} &= \hat{\theta}_1^T W_1(\zeta_1^{(0\sim m_1-1)}) + \Psi_1(\zeta_1^{(0\sim m_1-1)}, t), \\ \alpha_{12} &= 2L_1^2(\zeta_1^{(0\sim m_1-1)}, t)P_{1L}^T(A_1^{0\sim m_1-1})\zeta_1^{(0\sim m_1-1)}, \end{aligned} \tag{8.22}$$

$$\dot{\hat{\theta}}_1 = W_1(\zeta_1^{(0\sim m_1-1)}) P_{1L}^T (A_1^{0\sim m_1-1}) \zeta_1^{(0\sim m_1-1)} - \eta_1 \hat{\theta}_1, \tag{8.23}$$

where η_1 is a positive design parameter and $\hat{\theta}_1$ is the estimation of θ_1.

Substituting (8.21)–(8.23) into (8.20), we have

$$\zeta_1^{(m_1)} + A_1^{0\sim m_1-1} \zeta_1^{(0\sim m_1-1)} = \tilde{\theta}_1^T W_1(\zeta_1^{(0\sim m_1-1)}) + L_1(\zeta_1^{(0\sim m_1-1)}, t) h_2 \zeta_2$$
$$+ L_1(\zeta_1^{(0\sim m_1-1)}, t) h_2 \psi_{21} + \varepsilon_1 - \alpha_{12}, \tag{8.24}$$

where $\tilde{\theta}_1 = \theta_1 - \hat{\theta}_1$. And it can be rewritten as

$$\dot{\zeta}_1^{(0\sim m_1-1)} = \Phi(A_1^{0\sim m_1-1}) \zeta_1^{(0\sim m_1-1)}$$
$$+ \begin{bmatrix} 0 \\ \left(\begin{matrix} \tilde{\theta}_1^T W_1(\zeta_1^{(0\sim m_1-1)}) + L_1(\zeta_1^{(0\sim m_1-1)}, t) h_2 \zeta_2 \\ + L_1(\zeta_1^{(0\sim m_1-1)}, t) h_2 \psi_{21} + \varepsilon_1 - \alpha_{12} \end{matrix} \right) \end{bmatrix}. \tag{8.25}$$

Define the following Lyapunov function

$$V_1 = \frac{1}{2} \left(\zeta_1^{(0\sim m_1-1)} \right)^T P_1(A_1^{0\sim m_1-1}) \zeta_1^{(0\sim m_1-1)} + \frac{1}{2} \tilde{\theta}_1^T \tilde{\theta}_1. \tag{8.26}$$

Then taking derivative of V_1 yields

$$\dot{V}_1 = \frac{1}{2} \left(\zeta_1^{(0\sim m_1-1)} \right)^T \left(\Phi(A_1^{0\sim m_1-1})^T P_1(A_1^{0\sim m_1-1}) \right.$$
$$+ P_1(A_1^{0\sim m_1-1}) \Phi(A_1^{0\sim m_1-1}) \big)$$
$$\times \zeta_1^{(0\sim m_1-1)} - \dot{\hat{\theta}}_1^T \tilde{\theta}_1 + \left(\zeta_1^{(0\sim m_1-1)} \right)^T P_1(A_1^{0\sim m_1-1})$$
$$\times \begin{bmatrix} 0 \\ \left(\begin{matrix} \tilde{\theta}_1^T W_1(\zeta_1^{(0\sim m_1-1)}) + L_1(\zeta_1^{(0\sim m_1-1)}, t) h_2 \zeta_2 \\ + L_1(\zeta_1^{(0\sim m_1-1)}, t) h_2 \psi_{21} + \varepsilon_1 - \alpha_{12} \end{matrix} \right) \end{bmatrix}$$
$$\leq -\frac{1}{2} \mu_1 \left(\zeta_1^{(0\sim m_1-1)} \right)^T P_1(A_1^{0\sim m_1-1}) \zeta_1^{(0\sim m_1-1)} + \eta_1 \tilde{\theta}_1^T \hat{\theta}_1$$
$$+ \left(\zeta_1^{(0\sim m_1-1)} \right)^T P_{1L}(A_1^{0\sim m_1-1}) (L_1(\zeta_1^{(0\sim m_1-1)}, t) h_2 \zeta_2$$
$$+ L_1(\zeta_1^{(0\sim m_1-1)}, t) h_2 \psi_{21} + \varepsilon_1 - \alpha_{12}). \tag{8.27}$$

By employing Young's inequality, one has

$$
\left(\zeta_1^{(0\sim m_1-1)}\right)^T P_{1L}(A_1^{0\sim m_1-1})L_1(\zeta_1^{(0\sim m_1-1)},t)h_2\zeta_2
$$
$$
\leq L_1^2(\zeta_1^{(0\sim m_1-1)},t)\left(\zeta_1^{(0\sim m_1-1)}\right)^T P_{1L}(A_1^{0\sim m_1-1})P_{1L}^T(A_1^{0\sim m_1-1})\zeta_1^{(0\sim m_1-1)}
$$
$$
+\frac{1}{4\lambda_{\min}(P_2)}\left(\zeta_2^{(0\sim m_2-1)}\right)^T P_2(A_2^{0\sim m_2-1})\zeta_2^{(0\sim m_2-1)}, \tag{8.28}
$$

$$
\left(\zeta_1^{(0\sim m_1-1)}\right)^T P_{1L}(A_1^{0\sim m_1-1})L_1(\zeta_1^{(0\sim m_1-1)},t)h_2\psi_{21}
$$
$$
\leq L_1^2(\zeta_1^{(0\sim m_1-1)},t)\left(\zeta_1^{(0\sim m_1-1)}\right)^T P_{1L}(A_1^{0\sim m_1-1})P_{1L}^T(A_1^{0\sim m_1-1})\zeta_1^{(0\sim m_1-1)}
$$
$$
+\frac{1}{4\lambda_{\min}(P_{\psi2})}\psi_2^T P_{\psi2}\psi_2, \tag{8.29}
$$

$$
\left(\zeta_1^{(0\sim m_1-1)}\right)^T P_{1L}(A_1^{0\sim m_1-1})\varepsilon_1
$$
$$
\leq \left(\zeta_1^{(0\sim m_1-1)}\right)^T P_1(A_1^{0\sim m_1-1})\begin{bmatrix}0\\\varepsilon_1\end{bmatrix}
$$
$$
\leq \frac{\lambda_{\max}(P_1)}{\iota_1}\left(\zeta_1^{(0\sim m_1-1)}\right)^T P_1(A_1^{0\sim m_1-1})\zeta_1^{(0\sim m_1-1)}+\frac{\iota_1}{4}\bar{\varepsilon}_1^2, \tag{8.30}
$$

where $\lambda_{\max}(P_1)=\lambda_{\max}(P_1(A_1^{0\sim m_1-1}))$, $\lambda_{\min}(P_2)=\lambda_{\min}(P_2(A_2^{0\sim m_2-1}))$, and ι_1 is a positive designed constant.

Substituting (8.28)–(8.30) into (8.27), we have

$$
\dot{V}_1 \leq -\frac{1}{2}(\mu_1-2\frac{\lambda_{\max}(P_1)}{\iota_1}-\frac{1}{2\lambda_{\min}(P_1)})\left(\zeta_1^{(0\sim m_1-1)}\right)^T P_1(A_1^{0\sim m_1-1})\zeta_1^{(0\sim m_1-1)}
$$
$$
+\eta_1\tilde{\theta}_1^T\hat{\theta}_1+\frac{1}{4\lambda_{\min}(P_2)}\left(\zeta_2^{(0\sim m_2-1)}\right)^T P_2(A_2^{0\sim m_2-1})\zeta_2^{(0\sim m_2-1)}+\frac{\iota_1}{4}\bar{\varepsilon}_1^2
$$
$$
+\frac{1}{4\lambda_{\min}(P_{\psi2})}\psi_2^T P_{\psi2}\psi_2. \tag{8.31}
$$

Remark 8.4 To handle the influence of h_{i+1} in system dynamics, a novel command filtering architecture (8.14) is introduced. This design is partially inspired by methodologies in Zhao et al. (2019b), Liu et al. (2023b), and Liu et al. (2023c), serving dual purposes: resolving persistent virtual controller differentiation challenges and alleviating computational burdens during controller synthesis. Furthermore, the integration of RBF neural networks within the HOFA framework provides systematic approximation of nonlinear uncertainties, thereby expanding the methodological scope of the approach.

Step i $(i = 2, ..., n - 1)$: From (8.13), we have

$$\zeta_i^{(0\sim m_i)} = x_i^{(0\sim m_i)} - \alpha_{if}^{(0\sim m_i)}. \tag{8.32}$$

It follows that

$$
\begin{aligned}
\zeta_i^{(m_i)} &= F_i(\zeta_j^{(0\sim m_j-1)}|_{j=1\sim i}, t) + \Psi_i(\zeta_j^{(0\sim m_j-1)}|_{j=1\sim i}, t) \\
&\quad + L_i(\zeta_j^{(0\sim m_j-1)}|_{j=1\sim i}, t) \\
&\quad \times h_{i+1}(\zeta_{i+1} + \Psi_{i+1,1} + \frac{\alpha_i}{h_{i+1}}) - \alpha_{if}^{(m_i)},
\end{aligned}
\tag{8.33}
$$

in which

$$
\begin{aligned}
F_i(\zeta_j^{(0\sim m_j-1)}|_{j=1\sim i}, t) &= f_i(x_j^{(0\sim m_j-1)}|_{j=1\sim i}, t), \\
\Psi_i(\zeta_j^{(0\sim m_j-1)}|_{j=1\sim i}, t) &= \phi_i(x_j^{(0\sim m_j-1)}|_{j=1\sim i}, t), \\
L_i(\zeta_j^{(0\sim m_j-1)}|_{j=1\sim i}, t) &= g_i(x_j^{(0\sim m_j-1)}|_{j=1\sim i}, t).
\end{aligned}
$$

According to the neural approximation, $F_i(\zeta_j^{(0\sim m_j-1)}|_{j=1\sim i}, t)$ can be described by

$$F_i(\zeta_j^{(0\sim m_j-1)}|_{j=1\sim i}, t) = \theta_i^T W_i(\zeta_j^{(0\sim m_j-1)}|_{j=1\sim i}) + \varepsilon_i, \tag{8.34}$$

where ε_i is the approximation error and satisfies $\varepsilon_i < \bar{\varepsilon}_i$ with $\bar{\varepsilon}_i$ being a positive constant. Design

$$
\begin{aligned}
\alpha_i &= -L_i^{-1}(\zeta_j^{(0\sim m_j-1)}|_{j=1\sim i}, t)(A_i^{0\sim m_i-1}\zeta_i^{(0\sim m_i-1)} + \alpha_{i1} + \alpha_{i2}), \\
\alpha_{i1} &= \hat{\theta}_i^T W_i(\zeta_j^{(0\sim m_j-1)}|_{j=1\sim i}) + \Psi_i(\zeta_j^{(0\sim m_j-1)}|_{j=1\sim i}, t) - \alpha_{if}^{(m_i)}, \\
\alpha_{i2} &= 2L_i^2(\zeta_j^{(0\sim m_j-1)}|_{j=1\sim i}, t)P_{iL}^T(A_i^{0\sim m_i-1})\zeta_i^{(0\sim m_i-1)},
\end{aligned}
\tag{8.35}
$$

$$\dot{\hat{\theta}}_i = W_i(\zeta_j^{(0\sim m_j-1)}|_{j=1\sim i})P_{iL}^T(A_i^{0\sim m_i-1})\zeta_i^{(0\sim m_i-1)} - \eta_i\hat{\theta}_i, \tag{8.36}$$

where η_i is a positive design parameter, and $\hat{\theta}_i$ is the estimation of θ_i.

Substituting (8.34)–(8.36) into (8.33), we have

$$
\zeta_i^{(m_i)} + A_i^{0 \sim m_i-1} \zeta_i^{(0 \sim m_i-1)}
$$

$$
= \tilde{\theta}_i^T W_i(\zeta_j^{(0 \sim m_j-1)}|_{j=1 \sim i}) + L_i(\zeta_j^{(0 \sim m_j-1)}|_{j=1 \sim i}, t) h_{i+1} \zeta_{i+1}
$$

$$
+ L_i(\zeta_j^{(0 \sim m_j-1)}|_{j=1 \sim i}, t) h_{i+1} \psi_{i+1,1} + \varepsilon_i - \alpha_{i2}, \tag{8.37}
$$

where $\tilde{\theta}_i = \theta_i - \hat{\theta}_i$. The above equation can be rewritten as

$$
\dot{\zeta}_i^{(0 \sim m_i-1)} = \Phi(A_i^{0 \sim m_i-1}) \zeta_i^{(0 \sim m_i-1)}
$$

$$
+ \left[\begin{array}{c} 0 \\ \left(\begin{array}{l} L_i(\zeta_j^{(0 \sim m_j-1)}|_{j=1 \sim i}, t) h_{i+1} \zeta_{i+1} + L_i(\zeta_j^{(0 \sim m_j-1)}|_{j=1 \sim i}, t) \\ \times h_{i+1} \psi_{i+1,1} + \tilde{\theta}_i^T W_i(\zeta_j^{(0 \sim m_j-1)}|_{j=1 \sim i}) + \varepsilon_i - \alpha_{i2} \end{array} \right) \end{array} \right].
$$

$$
\tag{8.38}
$$

The Lyapunov function is chosen as

$$
V_i = V_{i-1} + \frac{1}{2} \left(\zeta_i^{(0 \sim m_i-1)} \right)^T P_i(A_i^{0 \sim m_i-1}) \zeta_i^{(0 \sim m_i-1)} + \frac{1}{2} \tilde{\theta}_i^T \tilde{\theta}_i, \tag{8.39}
$$

then taking derivative of V_i yields

$$
\dot{V}_i = \dot{V}_{i-1} + \frac{1}{2} \left(\zeta_i^{(0 \sim m_i-1)} \right)^T (\Phi(A_i^{0 \sim m_i-1})^T P_i(A_i^{0 \sim m_i-1})
$$

$$
+ P_i(A_i^{0 \sim m_i-1}) \Phi(A_i^{0 \sim m_i-1}))
$$

$$
\times \zeta_i^{(0 \sim m_i-1)} - \dot{\hat{\theta}}_i^T \tilde{\theta}_i + \left(\zeta_i^{(0 \sim m_i-1)} \right)^T P_i(A_i^{0 \sim m_i-1})
$$

$$
\times \left[\begin{array}{c} 0 \\ \left(\begin{array}{l} L_i(\zeta_j^{(0 \sim m_j-1)}|_{j=1 \sim i}, t) h_{i+1} \zeta_{i+1} + L_i(\zeta_j^{(0 \sim m_j-1)}|_{j=1 \sim i}, t) h_{i+1} \psi_{i+1,1} \\ + \tilde{\theta}_i^T W_i(\zeta_j^{(0 \sim m_j-1)}|_{j=1 \sim i}) + \varepsilon_i - \alpha_{i2} \end{array} \right) \end{array} \right]
$$

$$
\leq \dot{V}_{i-1} - \frac{1}{2} \mu_i \left(\zeta_i^{(0 \sim m_i-1)} \right)^T P_i(A_i^{0 \sim m_i-1}) \zeta_i^{(0 \sim m_i-1)} + \eta_i \tilde{\theta}_i^T \hat{\theta}_i
$$

$$
+ \left(\zeta_i^{(0 \sim m_i-1)} \right)^T P_{iL}(A_i^{0 \sim m_i-1})(L_i(\zeta_j^{(0 \sim m_j-1)}|_{j=1 \sim i}, t) h_{i+1} \zeta_{i+1}
$$

$$
+ L_i(\zeta_j^{(0 \sim m_j-1)}|_{j=1 \sim i}, t) h_{i+1} \psi_{i+1,1} + \varepsilon_i - \alpha_{i2}). \tag{8.40}
$$

By employing Young's inequality, one has

$$\left(\zeta_i^{(0\sim m_i-1)}\right)^T P_{iL}(A_i^{0\sim m_i-1})L_i(\zeta_j^{(0\sim m_j-1)}|_{j=1\sim i},t)h_{i+1}\zeta_{i+1}$$

$$\leq L_i^2(\zeta_j^{(0\sim m_j-1)}|_{j=1\sim i},t)\left(\zeta_i^{(0\sim m_i-1)}\right)^T P_{iL}(A_i^{0\sim m_i-1})P_{iL}^T(A_i^{0\sim 1})\zeta_i^{(0\sim m_i-1)}$$

$$+\frac{1}{4\lambda_{\min}(P_{i+1})}\left(\zeta_{i+1}^{(0\sim m_{i+1}-1)}\right)^T P_{i+1}(A_{i+1}^{0\sim m_{i+1}-1})\zeta_{i+1}^{(0\sim m_{i+1}-1)}, \qquad (8.41)$$

$$\left(\zeta_i^{(0\sim m_i-1)}\right)^T P_{iL}(A_i^{0\sim m_i-1})L_i(\zeta_j^{(0\sim m_j-1)}|_{j=1\sim i},t)h_{i+1}\psi_{i+1,1}$$

$$\leq L_i^2(\zeta_j^{(0\sim m_j-1)}|_{j=1\sim i},t)\left(\zeta_i^{(0\sim m_i-1)}\right)^T P_{iL}(A_i^{0\sim m_i-1})P_{iL}^T(A_i^{0\sim m_i-1})\zeta_i^{(0\sim m_i-1)}$$

$$+\frac{1}{4\lambda_{\min}(P_{\psi i+1})}\psi_{i+1}^T P_{\psi i+1}\psi_{i+1}, \qquad (8.42)$$

$$\left(\zeta_i^{(0\sim m_i-1)}\right)^T P_{iL}(A_i^{0\sim m_i-1})\varepsilon_i$$

$$\leq \left(\zeta_i^{(0\sim m_i-1)}\right)^T P_i(A_i^{0\sim m_i-1})\begin{bmatrix}0\\\varepsilon_i\end{bmatrix}$$

$$\leq \frac{\lambda_{\max}(P_i)}{\iota_i}\left(\zeta_i^{(0\sim m_i-1)}\right)^T P_i(A_i^{0\sim m_i-1})\zeta_i^{(0\sim m_i-1)}+\frac{\iota_i}{4}\bar{\varepsilon}_i^2, \qquad (8.43)$$

where $\lambda_{\max}(P_i)=\lambda_{\max}(P_i(A_i^{0\sim m_i-1}))$, $\lambda_{\min}(P_{i+1})=\lambda_{\min}(P_{i+1}(A_{i+1}^{0\sim m_{i+1}-1}))$, and ι_i is a positive designed constant.

Substituting (8.41)–(8.43) into (8.40), we have

$$\dot{V}_i \leq -\frac{1}{2}\sum_{j=1}^{i}(\mu_j-2\frac{\lambda_{\max}(P_j)}{\iota_j}-\frac{1}{2\lambda_{\min}(P_j)})\left(\zeta_j^{(0\sim m_j-1)}\right)^T$$

$$P_j(A_j^{0\sim m_j-1})\zeta_j^{(0\sim m_j-1)}$$

$$+\sum_{j=1}^{i}\eta_j\tilde{\theta}_j^T\hat{\theta}_j+\frac{1}{4\lambda_{\min}(P_{i+1})}\left(\zeta_{i+1}^{(0\sim m_{i+1}-1)}\right)^T P_{i+1}(A_{i+1}^{0\sim m_{i+1}-1})\zeta_{i+1}^{(0\sim m_{i+1}-1)}$$

$$+\sum_{j=1}^{i}\frac{1}{4\lambda_{\min}(P_{\psi j+1})}\psi_{j+1}^T P_{\psi j+1}\psi_{j+1}+\sum_{j=1}^{i}\frac{1}{4}\bar{\varepsilon}_j^2. \qquad (8.44)$$

Step n: Similar to this design process, we can finally obtain

$$\zeta_n^{(m_n)} = F_n(\zeta_j^{(0\sim m_j-1)}|_{j=1\sim n}, t) + \Psi_n(\zeta_j^{(0\sim m_j-1)}|_{j=1\sim n}, t)$$
$$+ L_n(\zeta_j^{(0\sim m_j-1)}|_{j=1\sim n}, t)u$$
$$- \alpha_{nf}^{(m_n)}, \tag{8.45}$$

in which

$$F_n(\zeta_j^{(0\sim m_j-1)}|_{j=1\sim n}, t) = f_n(x_j^{(0\sim m_j-1)}|_{j=1\sim n}, t),$$
$$\Psi_n(\zeta_j^{(0\sim m_j-1)}|_{j=1\sim n}, t) = \phi_n(x_j^{(0\sim m_j-1)}|_{j=1\sim n}, t),$$
$$L_n(\zeta_j^{(0\sim m_j-1)}|_{j=1\sim n}, t) = g_n(x_j^{(0\sim m_j-1)}|_{j=1\sim n}, t).$$

According to the neural approximation, $F_n(\zeta_j^{(0\sim m_j-1)}|_{j=1\sim n}, t)$ can be described by

$$F_n(\zeta_j^{(0\sim m_j-1)}|_{j=1\sim n}, t) = \theta_n^T W_n(\zeta_j^{(0\sim m_j-1)}|_{j=1\sim n}) + \varepsilon_n, \tag{8.46}$$

where ε_n is the approximation error and satisfies $\varepsilon_n < \bar{\varepsilon}_n$ with $\bar{\varepsilon}_n$ being a positive constant. The actual controller u and the adaptive law $\hat{\theta}_n$ are constructed as

$$u = -L_n^{-1}(\zeta_j^{(0\sim m_j-1)}|_{j=1\sim n}, t)(A_n^{0\sim m_n-1}\zeta_n^{(0\sim m_n-1)} + \alpha_{n1}),$$
$$\alpha_{n1} = \hat{\theta}_n^T W_n(\zeta_j^{(0\sim m_j-1)}|_{j=1\sim n}) + \Psi_n(\zeta_j^{(0\sim m_j-1)}|_{j=1\sim n}, t) - \alpha_{nf}^{(m_n)}, \tag{8.47}$$

$$\dot{\hat{\theta}}_n = W_n(\zeta_j^{(0\sim m_j-1)}|_{j=1\sim n})P_{nL}^T(A_n^{0\sim m_n-1})\zeta_n^{(0\sim m_n-1)} - \eta_n\hat{\theta}_n, \tag{8.48}$$

where η_n is a positive designed parameter and $\hat{\theta}_n$ is the estimation of θ_n. Substituting (8.46)–(8.48) into (8.45) yields

$$\zeta_n^{(m_n)} + A_n^{0\sim m_n-1}\zeta_n^{(0\sim m_n-1)} = \tilde{\theta}_n^T W_n(\zeta_j^{(0\sim m_j-1)}|_{j=1\sim n}) + \varepsilon_n, \tag{8.49}$$

where $\tilde{\theta}_n = \theta_n - \hat{\theta}_n$. Then (8.49) can be rewritten as

$$\dot{\zeta}_n^{(0\sim m_n-1)} = \Phi(A_n^{0\sim m_n-1})\zeta_n^{(0\sim m_n-1)} + \begin{bmatrix} 0 \\ \tilde{\theta}_n^T W_n(\zeta_j^{(0\sim m_j-1)}|_{j=1\sim n}) + \varepsilon_n \end{bmatrix}. \tag{8.50}$$

Define the Lyapunov function as

$$V_n = V_{n-1} + \frac{1}{2}\left(\zeta_n^{(0\sim m_n-1)}\right)^T P_n(A_n^{0\sim m_n-1})\zeta_n^{(0\sim m_n-1)} + \frac{1}{2}\tilde{\theta}_n^T\tilde{\theta}_n. \tag{8.51}$$

Then the derivative of V_n is

$$\dot{V}_n = \dot{V}_{n-1} + \frac{1}{2}\left(\zeta_n^{(0\sim m_n-1)}\right)^T (\Phi(A_n^{0\sim m_n-1})^T P_n(A_n^{0\sim m_n-1}) + P_n(A_n^{0\sim m_n-1})$$

$$\times \Phi(A_n^{0\sim m_n-1}))\zeta_n^{(0\sim m_n-1)} - \dot{\tilde{\theta}}_n^T\tilde{\theta}_n + \left(\zeta_n^{(0\sim m_n-1)}\right)^T P_n(A_n^{0\sim m_n-1})$$

$$\times \left[\begin{array}{c} 0 \\ \left(\tilde{\theta}_n^T W_i(\zeta_j^{(0\sim m_j-1)}|_{j=1\sim n}) + \varepsilon_n\right) \end{array}\right]$$

$$\leq \dot{V}_{n-1} - \frac{1}{2}\mu_n \left(\zeta_n^{(0\sim m_n-1)}\right)^T P_n(A_n^{0\sim m_n-1})\zeta_n^{(0\sim m_n-1)} + \eta_n\tilde{\theta}_n^T\hat{\theta}_n$$

$$+ \left(\zeta_n^{(0\sim m_n-1)}\right)^T P_{nL}(A_n^{0\sim m_n-1})\varepsilon_n. \tag{8.52}$$

By using Young's inequality, one has

$$\left(\zeta_n^{(0\sim m_n-1)}\right)^T P_{nL}(A_n^{0\sim m_n-1})\varepsilon_n$$

$$\leq \left(\zeta_n^{(0\sim m_n-1)}\right)^T P_n(A_n^{0\sim m_n-1})\left[\begin{array}{c} 0 \\ \varepsilon_n \end{array}\right]$$

$$\leq \frac{\lambda_{\max}(P_n)}{\iota_n}\left(\zeta_n^{(0\sim m_n-1)}\right)^T P_n(A_n^{0\sim m_n-1})\zeta_n^{(0\sim m_n-1)} + \frac{\iota_n}{4}\bar{\varepsilon}_n^2, \tag{8.53}$$

where $\lambda_{\max}(P_n) = \lambda_{\max}(P_n(A_n^{0\sim m_n-1}))$, and ι_n is a positive designed constant.
Substituting (8.53) into (8.52), we have

$$\dot{V}_n \leq -\frac{1}{2}\sum_{j=1}^{n}\mu_j^*\left(\zeta_j^{(0\sim m_j-1)}\right)^T P_j(A_j^{0\sim m_j-1})\zeta_j^{(0\sim m_j-1)} + \sum_{j=1}^{n}\eta_j\tilde{\theta}_j^T\hat{\theta}_j$$

$$+ \sum_{j=1}^{n}\frac{\iota_j}{4}\bar{\varepsilon}_j^2$$

$$+ \sum_{j=1}^{n-1}\frac{1}{4\lambda_{\min}(P_{\psi_{j+1}})}\psi_{j+1}^T P_{\psi_{j+1}}\psi_{j+1}, \tag{8.54}$$

where $\mu_i^* = (\mu_i - \frac{2\lambda_{\max}(P_i)}{\iota_i} - \frac{1}{2\lambda_{\min}(P_i)})$, $i = 1,\ldots,n$ are positive constants by selecting appropriate parameters.

Choose the Lyapunov function as

$$V = \sum_{i=2}^{n} V_{\psi_i} + V_n, \tag{8.55}$$

where $V_{\psi_i} = \frac{1}{2}\psi_i^T P_{\psi_i} \psi_i$.

Now we define the following compact set

$$\Omega_V = \{\sum_{i=1}^{n}((\zeta_i^{(0\sim m_i-1)})^T P_i(A_i^{0\sim m_i-1}) \times \zeta_i^{(0\sim m_i-1)} + \tilde{\theta}_i^T \tilde{\theta}_i)$$

$$+ \sum_{i=1}^{n-1} \psi_i^T P_{\psi_i} \psi_i \leq 2S\}$$

$$\subset \mathbb{R}^{dim(\Omega_V)}.$$

Therefore, within this compact set, there exist positive constants M_i such that $|((\frac{\alpha_{i-1}}{h_i})^{(m_i)} + \omega_{im_i}(\frac{\alpha_{i-1}}{h_i})^{(m_i-1)} + \cdots + \omega_{i2}(\frac{\alpha_{i-1}}{h_i})^{(1)})| \leq M_i, i = 2, \ldots, n$. Thus we have

$$\dot{V}_{\psi_i} = \frac{1}{2}\psi_i^T(A_{\psi_i}^T P_{\psi_i} + P_{\psi_i} A_{\psi_i})\psi_i + \psi_i^T P_{\psi_i} P_{\psi_i}^T \psi_i$$

$$+ \frac{1}{4}((\frac{\alpha_{i-1}}{h_i})^{(m_i)} + \omega_{im_i}(\frac{\alpha_{i-1}}{h_i})^{(m_i-1)}$$

$$+ \cdots + \omega_{i2}(\frac{\alpha_{i-1}}{h_i})^{(1)})^2$$

$$\leq -\frac{1}{2}(\mu_{\psi_i} - 2\lambda_{\max}(P_{\psi_i}))\psi_i^T P_{\psi_i} \psi_i + \frac{1}{4}M_i^2. \tag{8.56}$$

Then it follows

$$\dot{V} \leq -\frac{1}{2}\sum_{j=1}^{n}\mu_j^*\left(\zeta_j^{(0\sim m_j-1)}\right)^T P_j(A_j^{0\sim m_j-1})\zeta_j^{(0\sim m_j-1)}$$

$$-\frac{1}{2}\sum_{j=1}^{n-1}(\mu_{\psi_{j+1}} - 2\lambda_{\max}(P_{\psi_{j+1}}) - \frac{1}{2\lambda_{\min}(P_{\psi_{j+1}})})\psi_{j+1}^T P_{\psi_{j+1}} \psi_{j+1}$$

$$+ \sum_{j=1}^{n}\eta_j\tilde{\theta}_j^T \hat{\theta}_j + \sum_{j=1}^{n}\frac{\iota_j}{4}\bar{\varepsilon}_j^2 + \frac{1}{4}\sum_{j=1}^{n-1}M_{j+1}^2. \tag{8.57}$$

Due to

$$\eta_i \tilde{\theta}_i^T \hat{\theta}_i \le -\frac{1}{2}\eta_i \tilde{\theta}_i^T \tilde{\theta}_i + \frac{1}{2}\eta_i \left\| \theta_i \right\|^2, \tag{8.58}$$

we have

$$\dot{V} \le -\frac{1}{2}\sum_{j=1}^{n} \mu_j^* \left(\zeta_j^{(0\sim m_j-1)} \right)^T P_j(A_j^{0\sim m_j-1}) \zeta_j^{(0\sim m_j-1)} - \frac{1}{2}\sum_{j=1}^{n} \eta_j \tilde{\theta}_j^T \tilde{\theta}_j$$

$$-\frac{1}{2}\sum_{j=1}^{n-1}(\mu_{\psi_{j+1}} - 2\lambda_{\max}(P_{\psi_{j+1}})$$

$$-\frac{1}{2\lambda_{\min}(P_{\psi_{j+1}})}) \psi_{j+1}^T P_{\psi_{j+1}} \psi_{j+1} + \frac{1}{4}\sum_{j=1}^{n-1} M_{j+1}^2$$

$$+\frac{1}{2}\sum_{j=1}^{n} \eta_j \left\| \theta_j \right\|^2 + \sum_{j=1}^{n} \frac{\iota_j}{4}\bar{\varepsilon}_j^2$$

$$\le -CV + D, \tag{8.59}$$

where $D = \frac{1}{4}\sum_{j=1}^{n-1} M_{j+1}^2 + \frac{1}{2}\sum_{j=1}^{n} \eta_j \left\| \theta_j \right\|^2 + \sum_{j=1}^{n} \frac{\iota_j}{4}\bar{\varepsilon}_j^2$, $C = \min_{1\le j\le n-1}\{\mu_j^*, \mu_n^*,$
$\eta_j, \eta_n, (\mu_{\psi_{j+1}} - 2\lambda_{\max}(P_{\psi_{j+1}}) - \frac{1}{2\lambda_{\min}(P_{\psi_{j+1}})})\}$ with $(\mu_{\psi_{j+1}} - 2\lambda_{\max}(P_{\psi_{j+1}}) - \frac{1}{2\lambda_{\min}(P_{\psi_{j+1}})})$, $j = 1, \ldots, n-1$ being positive constants by selecting appropriate parameters. It implies that $\dot{V} \le 0$ on $V = S$ when $C > \frac{D}{S}$. Therefore, $V \le S$ is an invariant set, that is, if $V(0) \le S$, then $V(t) \le S$ for $t \ge 0$, which ensures the boundness of signals $\zeta_i^{(0\sim m_i-1)}$, $\hat{\theta}_i$ ($i = 1, \ldots, n$), and ψ_i ($i = 2, \ldots, n$).

8.4 Stability Analysis Under Multi-Variable Constraint

Based on the controller design process, we can generalize the following theorem and analyze the stability of the HOSF fully actuated nonlinear system under the proposed control strategy.

Theorem 8.1 *Consider the uncertain HOSF fully actuated nonlinear system (8.1). The implementation of virtual controllers (8.22), (8.35), actual controller (8.47), and adaptive laws (8.23), (8.36) and (8.48) guarantees the boundedness of all system signals while the system states maintain within the prescribed constraints.*

Proof By multiplying (8.59) by e^{Ct}, one obtains $d(V(t)e^{Ct})/dt \le De^{Ct}$. Integrating this inequality over the interval $[0, t]$ results in

$$V(t) \le \left(V(0) - \frac{D}{C} \right) e^{-Ct} + \frac{D}{C}, \tag{8.60}$$

which implies $V(t)$ is bounded over the time period $[0, \infty)$. Furthermore, as $t \to \infty$, $V \to \frac{D}{C}$. Since

$$V = \frac{1}{2} \sum_{i=1}^{n} (\zeta_i^{(0 \sim m_i - 1)})^T P_i (A_i^{0 \sim m_i - 1}) \zeta_i^{(0 \sim m_i - 1)} + \frac{1}{2} \sum_{i=1}^{n} \tilde{\theta}_i^T \tilde{\theta}_i + \frac{1}{2} \sum_{i=1}^{n-1} \psi_i^T P_{\psi_i} \psi_i, , \tag{8.61}$$

it follows that $\zeta_i^{(0 \sim m_i - 1)}$ eventually converges into

$$\Theta_{(C,D)}(0) = \{ \zeta_i^{(0 \sim m_i - 1)} | \| \zeta_i^{(0 \sim m_i - 1)} \|_{P_i}^2 \le 2 \frac{D}{C} \}. \tag{8.62}$$

From (8.60), the boundedness of ζ_i implies $|q_i(t)| < \rho_i(t)$ and $|q_i(t)| < \rho_{iT_f}$ for $t \le T_f$, demonstrating state confinement within prescribed boundaries and convergence to the invariant region within time T_f regardless of initial conditions. Given the bounded nature of $\hat{\theta}_1$, $\zeta_1^{(0 \sim m_1 - 1)}$, and ψ_2, Eqs. (8.14)–(8.16) ensure bounded α_1, $\alpha_{2,f}$, $\dot{\alpha}_{2f}$, ..., $\alpha_{2f}^{(m_2)}$. By extension, boundedness propagates through α_i, $\alpha_{i+1,f}$, $\dot{\alpha}_{i+1,f}$, ..., $\alpha_{i+1,f}^{(m_{i+1})}$ ($i = 2, ..., n-1$), guaranteeing bounded control input u. Adjustable parameters ρ_{iT_f} and T_f enable performance customization while maintaining signal boundedness throughout the closed-loop system.

Remark 8.5 For the investigated uncertain HOSF system (1), conventional constrained control methodologies typically require system decomposition into cascaded first-order subsystems, introducing significant computational burdens. This work introduces an innovative integration of the HOFA framework with PPTC, achieving dual objectives: reduced computational steps and enhanced transient performance. Crucially, the proposed methodology eliminates initialization constraints on system states, thereby relaxing operational requirements.

Remark 8.6 Matrix selections for $A_i^{0 \sim m_i - 1}$ and $P_i(A_i^{0 \sim m_i - 1})$ must comply with Proposition 8.1 and Lemmas 2.5 and 2.6, respectively. A sufficient configuration involves selecting diagonal matrix F with negative entries. For complex eigenvalue requirements, incorporate block matrices of the form:

$$\begin{bmatrix} -a & -b \\ b & -a \end{bmatrix}$$

where $a, b > 0$, as diagonal components of F. Following Proposition 8.1, $A_i^{0 \sim m_i - 1}$ can be systematically constructed. System performance optimization strategies include:

- Modifying performance function $\rho_i(t)$ and compound mapping $h_i(\gamma_i)$ to accelerate convergence and tighten prescribed bounds
- Increasing gain parameters η_i, μ_{ψ_i}, and stability threshold μ_i
- Reducing residual term D's design parameter magnitude.

8.5   Numerical Simulations Demonstrating Multi-Variable Constrained Performance

To verify the effectiveness and efficiency of our control strategy, the following practical electromechanical system as described in Liu et al. (2023b) is simulated

$$\begin{cases} M\ddot{q} + B\dot{q} + N\sin(q) = I, \\ L\dot{I} = V_\varepsilon - RI - K_B\dot{q}, \end{cases} \tag{8.63}$$

where M, N, and B are defined as

$$M = \frac{J}{K_\tau} + \frac{mL_0^2}{3K_\tau} + \frac{M_0 L_0^2}{K_\tau} + \frac{2M_0 R_0^2}{5K_\tau}, \tag{8.64}$$

$$N = \frac{M_0 L_0 G}{2K_\tau} + \frac{2M_0 L_0 G}{K_\tau}, \quad B = \frac{B_0^2}{K_\tau}. \tag{8.65}$$

The variables J, m, M_0, L_0, R_0, G, B_0, $q(t)$, $I(t)$, and K_τ represent the rotor inertia, link mass, load mass, link length, load radius, gravity coefficient, joint viscous friction coefficient, motor angular position, motor armature current, and electromechanical conversion coefficient, respectively. Additionally, L, V_ε, R, and K_B denote its inductance, voltage, resistance, and back EMF coefficient, respectively. It is evident that each subsystem in the above practical electromechanical system is a fully actuated system with the characteristics of fully actuated, given that both M and L are non-zero. By treating $q_1 = q$, $q_2 = I$, and $u = V_\varepsilon$, the physical system described above can be converted into the same form as (8.1), thereby facilitating the application of HOFA system approach. The resulting transformed system is

$$\begin{cases} q_1^{(m_1)} = p_1 q_2 + f_1(q_1^{(m_1 - 1)}), \\ q_2^{(m_2)} = p_2 u + f_2(q_j^{(m_j - 1)})|_{j=1 \sim 2}, \end{cases} \tag{8.66}$$

where $m_1 = 2$, $m_2 = 1$, $p_1 = \frac{1}{M}$, $p_2 = \frac{1}{L}$, $f_1 = \frac{1}{M}(B\dot{q} + N\sin(q))$, $f_2 = \frac{1}{L}(RI - K_B\dot{q})$. We use the parameter values $J = 1.625 \times 10^{-3}$ kg · m^2, $m = 0.506$ kg, $R_0 = 0.023$ m, $M_0 = 0.434$ kg, $L_0 = 0.305$ m, $B_0 = 16.25 \times 10^{-3}$ N · m · s/rad, $L = 15$ H, $R = 5.0\,\Omega$, and $K_\tau = K_B = 0.90$ N · m/A for this simulation.

The parameters of the prescribed constraints chosen as $\rho_{1T_f} = 0.03$, $\rho_{2T_f} = 0.3$, $T_f = 2$ s, $p = 2$, $c_1 = c_2 = 50$, $a_1 = 3.045$, and $a_2 = 24.5$. Then, design the corresponding controllers and adaptive laws as

$$\alpha_1 = -L_1^{-1}(\zeta_1^{(0\sim1)}, t)(A_1^{0\sim1}\zeta_1^{(0\sim1)} + \hat{\theta}_1^T W_1(\zeta_1^{(0\sim1)}) + \Psi_1(\zeta_1^{(0\sim1)}, t)$$
$$+ 2L_1^2(\zeta_1^{(0\sim1)}, t)P_{1L}^T(A_1^{0\sim1})\zeta_1^{(0\sim1)}), \tag{8.67}$$

$$u = -L_2^{-1}(\zeta_j^{(0\sim m_j)}|_{j=1\sim2}, t)(A_2\zeta_2 + \hat{\theta}_2^T W_2(\zeta_j^{(0\sim m_j)}|_{j=1\sim2})$$
$$+ \Psi_2(\zeta_j^{(0\sim m_j)}|_{j=1\sim2}, t) - \alpha_{2,f}^{(2)}), \tag{8.68}$$

$$\dot{\hat{\theta}}_1 = W_1(\zeta_1^{(0\sim m_1)})P_{1L}^T(A_1^{0\sim1})\zeta_1^{(0\sim1)} - \eta_1\hat{\theta}_1, \tag{8.69}$$

$$\dot{\hat{\theta}}_2 = W_2(\zeta_j^{(0\sim m_j)}|_{j=1\sim2})P_{2L}^T(A_2)\zeta_2 - \eta_2\hat{\theta}_2, \tag{8.70}$$

where $A_2 = 6$, $P_2 = P_{2L} = 1$, $\eta_1 = \eta_2 = 1.5$.

The solution of $A_1^{0\sim1}$ and $P_{1L}^T(A_1^{0\sim1})$ can be derived based on Proposition 8.1 and Lemmas 2.5–2.6. Choose matrix F as

$$F = \begin{bmatrix} -3 & -1 \\ 1 & -3 \end{bmatrix}, \tag{8.71}$$

whose eigenvalues are $-3 \pm j$, and $Z = [1, 1]$. Then we obtain

$$V = \begin{bmatrix} Z \\ ZF \end{bmatrix} = \begin{bmatrix} 1 & 1 \\ -2 & -4 \end{bmatrix}, \tag{8.72}$$

and $A_1^{0\sim1} = -ZF^2V^{-1} = \begin{bmatrix} 10 & 6 \end{bmatrix}$. We can also derive that

$$\Phi(A_1^{0\sim1}) = VFV^{-1} = \begin{bmatrix} 0 & 1 \\ -10 & -6 \end{bmatrix}, \tag{8.73}$$

which has the eigenvalues $-3 \pm j$. Considering the Lyapunov equation

$$(\Phi(A_1^{0\sim1}) + 2I)^T P_1(A_1^{0\sim1}) + P_1(A_1^{0\sim1})(\Phi(A_1^{0\sim1}) + 2I) = -I, \tag{8.74}$$

Fig. 8.1 The responses of states q_1 and q_2 under the proposed control strategy and the comparative control strategy. (**a**) The response of q_1. (**b**) The response of q_2

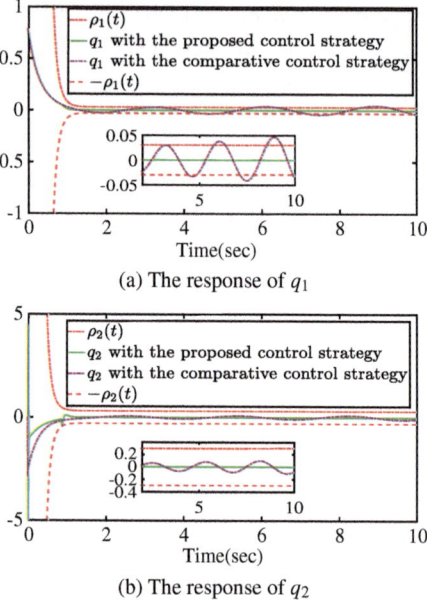

(a) The response of q_1

(b) The response of q_2

we can see that $\Phi^T(A_1^{0\sim1})P_1(A_1^{0\sim1}) + P_1(A_1^{0\sim1})\Phi(A_1^{0\sim1}) \leq -\mu_1 P_1(A_1^{0\sim1})$ holds with $\mu_1 = 4$. By solving the above equation, we obtain

$$P_1(A_1^{0\sim1}) = \begin{bmatrix} 14.75 & 3 \\ 3 & 0.875 \end{bmatrix},$$

which means $P_{1L}^T(A_i^{0\sim1}) = \begin{bmatrix} 3 & 0.875 \end{bmatrix}$. Based on the selections for the matrices $A_1^{0\sim1}$, $P_1(A_1^{0\sim1})$, A_2, and P_2, we can obtain $\mu_1 = 4$ and $\mu_2 = 5$. From the definition of μ_i^* in (58) and the values of $\lambda_{\min}(P_i)$, $\lambda_{\min}(P_i)$, and $\lambda_{\max}(P_i)$, it is derived that $\mu_i^* > 0$ by selecting appropriate parameters, which achieves the bounded stability of the system.

The simulation employs initial conditions $q_1(0) = \pi/4$, $\dot{q}_1(0) = 0$, $q_2(0) = 0$, and $\hat{\theta}_i(0) = [0.5, 0.4, 0.3, 0.2, 0.1]^T$ ($i = 1, 2$). Figures 8.1, 8.2, 8.3, and 8.4 validate the effectiveness of the proposed control scheme. Figure 8.1 confirms that all system states remain strictly bounded by prescribed performance functions while converging to the invariant region within the predefined time T_f. To emphasize the advantage of our method, comparative simulations against the PPTC-free strategy in Liu et al. (2023b) were conducted under identical parameters. Results in Fig. 8.1a, b demonstrate superior convergence speed and constraint compliance of our approach. Figure 8.2 reveals reduced $\dot{q}_1$ oscillation amplitudes with our controller. The neural network adaptive laws $\hat{\theta}_1$ and $\hat{\theta}_2$ exhibit stable convergence in Fig. 8.3. Figure 8.4a, b respectively display bounded virtual control α_1 and actual control u, confirming system stability.

Fig. 8.2 The responses of state $\dot{q}_1$ under the proposed control strategy and the comparative control strategy. (**a**) The response of $\dot{q}_1$. (**b**) The response of $\dot{q}_1$

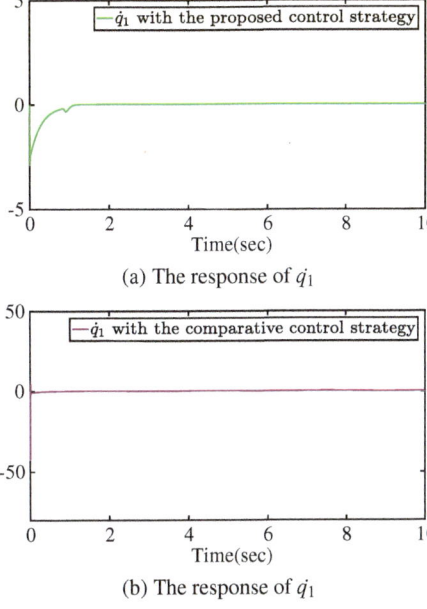

(a) The response of $\dot{q}_1$

(b) The response of $\dot{q}_1$

Fig. 8.3 The responses of the adaptive laws $\hat{\theta}_1$ and $\hat{\theta}_2$. (**a**) The response of $\hat{\theta}_1$. (**b**) The response of $\hat{\theta}_2$

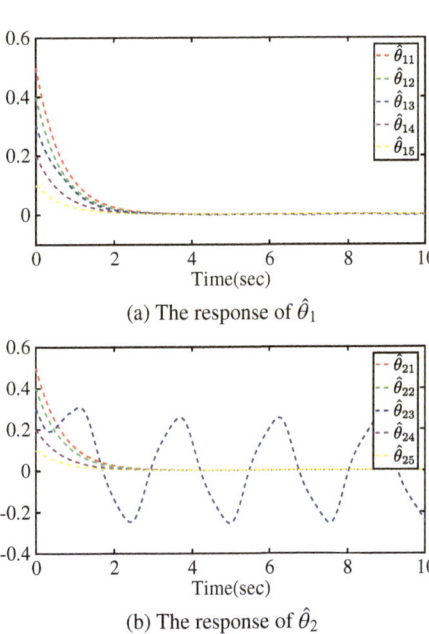

(a) The response of $\hat{\theta}_1$

(b) The response of $\hat{\theta}_2$

Fig. 8.4 The responses of virtual control input α_1 and actual control input u. (**a**) The response of α_1. (**b**) The response of u

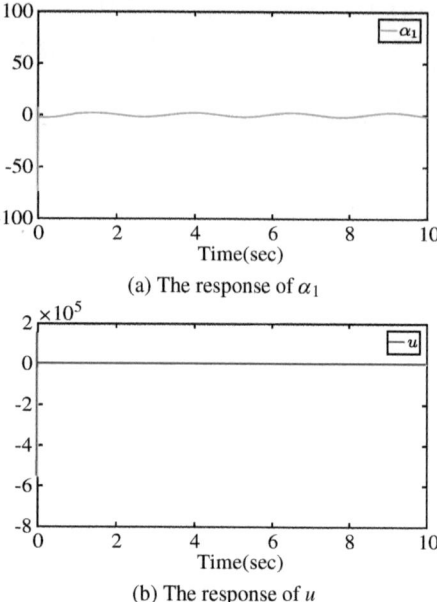

(a) The response of α_1

(b) The response of u

Notably, achieving state confinement within 1 s (Fig. 8.1) necessitates substantial control energy—a practically reasonable trade-off. These comprehensive results substantiate the control scheme's capability to fulfill the specified objectives.

8.6   Notes and References

In this chapter, a novel multi-variable constrained control based on the HOFA system approach is proposed for uncertain HOSF fully actuated constrained nonlinear systems. The key innovation resides in integrating command filtered control and RBF neural network techniques into the HOFA system approach control design while tackling state constraints for the mentioned nonlinear systems. This control strategy efficiently addresses the unknown nonlinear functions, facilitating system state convergence to the prescribed invariant region within a predetermined time. Finally, the scheme's efficacy is demonstrated through a simulation study.

Chapter 9
Adaptive Tracking Error-Constraint Control of Fully Actuated Unmanned Underwater Vehicle System

This chapter focuses on the adaptive trajectory tracking of unmanned underwater vehicles (UUVs) using a fully actuated system approach. UUVs face challenges such as hydrodynamic uncertainties, external disturbances, and actuator limitations. We propose an adaptive tracking error-constraint control strategy that incorporates normalized and barrier functions, combined with adaptive RBF neural networks, to handle these challenges. The proposed method ensures tracking errors remain within acceptable bounds, even under uncertainties.

9.1 Overview

UUVs find extensive applications in military operations, underwater resource exploration, underwater search missions, and related domains (Mohan & Kim 2015; Martin & Whitcomb 2018; Weisler et al. 2018; Yang et al. 2024; Wang 2021; Liu et al. 2021a). Successful execution of these tasks necessitates precise UUV trajectory tracking as emphasized in Wang (2021). However, the inherent nonlinearity and strong coupling within UUV control models significantly increase the complexity of system modeling. Furthermore, the complex underwater environment subjects UUVs to unpredictable influences from external disturbances, uncertain inertia parameters, and unobservable motion states. Leveraging the function approximation capability of neural networks (NNs), researchers have extensively explored NN-based methods to mitigate UUV model uncertainties. For instance, Shojaei and Arefi (2015) applied a NN approximation scheme to compensate for unknown model parameters and external disturbances caused by ocean currents and waves, achieving uniform ultimate boundedness of tracking errors. In Liu et al. (2024b), an adaptive trajectory tracking controller incorporating three adaptive laws was developed for UUVs with uncertain dynamics and disturbances, utilizing the approximation properties of command filters and NNs. Cui et al. (2017) addressed the trajectory tracking problem for a horizontal-plane UUV by integrating two NNs (a critic NN

© The Author(s) 2026
C. Hua et al., *Adaptive Constrained Control for High-Order Fully Actuated Nonlinear Systems*, https://doi.org/10.1007/978-981-95-0962-1_9

and an action NN) into an adaptive controller design. Additionally, dead-zone limitations in UUV propulsion systems can cause inaccuracies in controlling position, orientation, or velocity. Shen et al. (2017) investigated the output consensus problem for UUVs subject to saturation and dead-zone input constraints within a directed graph framework. Wang et al. (2023c) proposed an adaptive command filtered backstepping approach with dead-zone compensation to handle unknown actuator dead-zone nonlinearity. It is noteworthy that opportunities for enhancement in precise and stable tracking performance persist when UUVs operate in underwater task scenarios.

Operating UUVs within oceanic environmental constraints necessitates addressing various restriction issues to satisfy system performance and operational safety requirements (Li et al. 2022a; Chen & Hua 2022; Liu et al. 2021b, 2022b; Pan et al. 2024; Hua et al. 2022). These constraints are typically categorized into three primary forms: input constraints, state constraints, and output constraints. Barrier functions represent the principal systematic methodology for handling such constrained control challenges in UUV applications. Liu et al. (2021b) introduced an innovative adaptive region tracking control approach employing nonlinear error transformation via prescribed performance barrier function (PPBF). Implementation of PPBF in Liu et al. (2022b) ensured tracking errors remained within predefined boundaries, enhancing UUV safety during obstacle avoidance maneuvers. Hua et al. (2022) addressed concurrent velocity and position limitations through an adaptive position-velocity constrained tracking controller design. To improve UUV tracking precision and bandwidth utilization, Shi et al. (2023) established a prescribed performance sectionalized event-triggered framework. Liang et al. (2021) developed a finite-time velocity-observer based adaptive output feedback controller for UUV formation control, confining all output states within prescribed performance bounds. These control strategies effectively prevent underwater obstacle collisions. Nevertheless, the PPBF-based methodology inherently exhibits initial condition dependence and yields semiglobal outcomes from a global perspective.

Recently, the FAS methodology has gained substantial research interest due to its proven efficacy and simplicity in addressing control complexities characteristic of second-order or higher-order nonlinear systems (Duan 2020a, 2021b,c,d,e; Duan & Zhao 2022; Zhang et al. 2023e; Duan 2021f; Zhang et al. 2023b,c). This innovative approach has been rigorously tested and demonstrated effectiveness across multiple domains such as spacecraft systems (Duan & Zhao 2022; Zhang et al. 2023b), multiagent robotic systems (Zhang et al. 2023c), turbofan engine systems Peng et al. (2023), and related applications. Duan and Zhao (2022) resolved six-degree-of-freedom spacecraft position and attitude tracking control by developing an extended state observer for comprehensive disturbance estimation, accounting for uncertainties from external disturbances and dynamic model variations. Zhang et al. (2023b) tackled output tracking in discrete-time HOFA systems with implementation on air-bearing spacecraft simulators. Cooperative control of HOFA networked multiagent systems featuring communication delays between network nodes, sensors, and actuators was examined in Zhang et al. (2023c). Peng et al. (2023) devised a compound controller accomplishing multivariable control decoupling subject to aero-engine constraints. Through FAS methodology, UUV models can be transformed into fully actuated unmanned underwater vehicle systems (FAUUVs), enabling closed-loop systems with arbitrarily assignable eigenstructures via controller implementation. However, practical FAUUVs models inevitably encounter inherent model uncertainties and unknown dead-zone inputs, with minimal literature applying NNs techniques to FAS with uncertain dynamics. Consequently, integrating FAS methodology with NNs techniques to resolve uncertain dynamic models and unknown dead-zone inputs in UUVs constitutes a critical research challenge.

This chapter presents the design of an innovative adaptive trajectory tracking error constraint controller, accounting for the effects of unknown hydrodynamics, external disturbances, and unknown dead-zone input. The contributions of this research are outlined as follows:

1. In contrast to Liu et al. (2021b, 2022b), a novel error constraint fully actuated systems (ECFAS) approach that utilizes the tracking error dependent normalized function is presented. Based on the ECFAS approach, the UUVs model is transformed into FAUUVs model, which not only can accurately describe the UUVs model but also simplify the controller design process.
2. Utilizing a time-varying scaling function, a specific PPBF is introduced that is agnostic to the initial conditions of UUVs. This combination of the ECFAS approach and PPBF results in the transformation of the UUVs model into the FAUUVs model.
3. The inherent uncertainties within hydrodynamics are considered in the model of UUVs. In order to surmount the intricate challenge posed by unknown hydrodynamic characteristics in the FAUUVs model, the utilization of NNs technique emerges as a promising avenue, fostering an enhanced understanding of the intricate system.

9.2 System Formulation

The control objective is to design an adaptive state-feedback control scheme for the uncertain fully actuated system (9.6) such that the output of closed-loop systems tracks the reference signal $\eta_r = [\eta_{r1}, \eta_{r2}, \ldots, \eta_{r6}]^T \in \mathbb{R}^6$ asymptotically, while guaranteeing that the tracking error consistently remains within a prescribed boundary, regardless of the initial conditions.

9.2.1 Fully Actuated Unmanned Underwater Vehicle System

We consider the following multiple input multiple output (MIMO) UUVs systems

$$\begin{cases} \dot{\eta} = J(\eta)\mu, \\ M\dot{\mu} + C(\mu)\mu + D(\mu) + G(\eta) = D_z(\tau) + \tau_d, \end{cases} \tag{9.1}$$

Fig. 9.1 UUVs with
reference frames

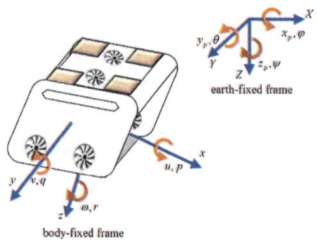

where $\eta = [x_p, y_p, z_p, \varphi, \theta, \psi]^T \in \mathbb{R}^6$ denotes the position and orientation
of UUVs, with elements associated with sway, surge, heave, roll, pitch and yaw
motions. $\mu = [u, v, \omega, p, q, r]^T \in \mathbb{R}^6$ represents the linear/angular velocity vector.
The symmetric positive-definite inertia matrix $M \in \mathbb{R}^{6 \times 6}$ incorporates added mass
effects. The composite Coriolis matrix $C(\mu) = C_0 + C_d \in \mathbb{R}^{6 \times 6}$ and restoring
vector $G(\eta) = G_0 + G_d \in \mathbb{R}^6$ consist of nominal components (C_0, G_0) and
uncertainty terms (C_d, G_d). The damping effects are characterized by $D(\mu) \in \mathbb{R}^6$,
while $\tau = [\tau_1, \tau_2, \tau_3, \tau_4, \tau_5, \tau_6]^T \in \mathbb{R}^6$ denotes the control input vector subject to
time-varying disturbances $\tau_d \in \mathbb{R}^6$, D_z is the Dead-Zone input.

The transformation matrix $J(\eta) \in \mathbb{R}^{6 \times 6}$ characterizes the kinematic relationship
between earth-fixed and body-fixed coordinate frames (see Fig. 9.1), given by

$$J(\eta) = \begin{bmatrix} c\psi c\theta & c\psi s\phi s\theta - c\phi s\psi & s\phi s\psi + c\phi c\psi s\theta & 0 & 0 & 0 \\ c\theta s\psi & c\phi c\psi + s\phi s\psi s\theta & c\phi s\psi s\theta - c\psi s\phi & 0 & 0 & 0 \\ -s\theta & c\theta s\phi & c\phi c\theta & 0 & 0 & 0 \\ 0 & 0 & 0 & 1 & s\phi t\theta & c\phi t\theta \\ 0 & 0 & 0 & 0 & c\phi & -s\phi \\ 0 & 0 & 0 & 0 & \frac{s\phi}{c\theta} & \frac{c\phi}{c\theta} \end{bmatrix}, \qquad (9.2)$$

where $s(*), c(*), t(*)$ denote $\sin(*), \cos(*)$ and $\tan(*)$, respectively.

9.2.2 Nonlinear Dead-Zone Characteristic

$D_{zi}(\tau_i), i = 1, 2, \ldots, 6$ is an asymmetric dead-zone nonlinear characteristic and it
is defined in the following form:

$$D_{zi}(\tau_i) = \begin{cases} m_{ri}(t)(\tau_i - b_{ri}), & \tau_i \geq b_{ri}, \\ 0, & b_{li} < \tau_i < b_{ri}, \\ m_{li}(t)(\tau_i - b_{li}), & \tau_i \leq b_{li}, \end{cases} \qquad (9.3)$$

where $m_{li}(t)$ and $m_{ri}(t)$ denote the time-varying left/right slopes of the dead-
zone nonlinearity. The breakpoint parameters $b_{li} < 0$ (left) and $b_{ri} > 0$ (right)
define the input discontinuity thresholds, with $b_{ri} - b_{li}$ specifying the dead-zone

Fig. 9.2 Illustration of the dead-zone characteristic

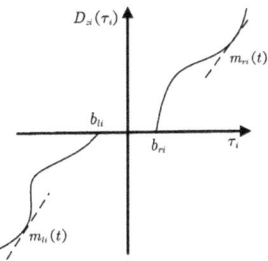

width. As illustrated in Fig. 9.2, the asymmetric dead-zone profile is schematically characterized.

The dead-zone input characteristic can be expressed in the following form:

$$D_z(\tau) = \begin{bmatrix} m_1(t) \\ & \ddots \\ & & m_6(t) \end{bmatrix} \tau + \begin{bmatrix} d_1(t) \\ \vdots \\ d_6(t) \end{bmatrix} = m(t)\tau + d(t), \tag{9.4}$$

with

$$m_i(t) = \begin{cases} m_{ri}(t) & \tau_i > 0 \\ m_{li}(t) & \tau_i \leq 0 \end{cases}, d_i(t) = \begin{cases} -m_{ri}(t)b_{ri}, & \tau_i \geq b_{ri}, \\ -m_i(t)\tau_i, & b_{li} < \tau_i < b_{ri}, \\ -m_{li}(t)b_{li}, & \tau_i \leq b_l. \end{cases}$$

Then, a system transformation can be described as

$$\begin{cases} \eta = z, \\ \mu = J^{-1}(z)\dot{z}. \end{cases} \tag{9.5}$$

Combining (9.1) and (9.2), we get a FAUUVs model:

$$\ddot{z} = f\left(z^{(0\sim1)}\right) + g(z^{(0\sim1)})\tau + F + \tau_\mu, \tag{9.6}$$

where $f(z^{(0\sim1)}) = -J(z)[\dot{J}^{-1}(z)\dot{z} + M^{-1}(C_0(\dot{z})\dot{z} + G_0(z))] \in \mathbb{R}^6$ is a smooth vector function, $g(z^{(0\sim1)}) = J(z)M^{-1}m(t) \in \mathbb{R}^{6\times6}$ is the unknown positive definite control matrix function, $F = -J(z)M^{-1}(C_d(\dot{z})\dot{z} + D(\dot{z}) + G_d(z)) \in \mathbb{R}^6$ is the lumped model uncertainties needs to be estimated. In particular, $\tau_\mu = J(z)M^{-1}(\tau_d + d(t)) \in \mathbb{R}^6$ is the disturbance variable with the time-varying external disturbances vector τ_d.

9.2.3 Preliminaries

To address the control objectives, essential assumptions and fundamental lemmas are established as prerequisites.

Assumption 9.1 The desired trajectory η_r and its derivatives up to n-th are known, bounded, and piecewise continuous.

Assumption 9.2 The roll angle φ, the pitch angle θ and yaw angle ψ are confined to the interval $(-90°, 90°)$.

Assumption 9.3 For the dead zone characteristic $D_{zi}(\tau_i)$, $i = 1, 2, \ldots, 6$, assume that $b_{ri} > 0$ and $b_{li} < 0$ are unknown constants, and $m_{ri}(t)$ and $m_{li}(t)$ are unknown positive bounded functions.

Assumption 9.4 The external time-varying disturbances τ_d are always bounded.

Remark 9.1 Stemming from mechanical reliability and operational safety constraints, the roll angle φ, pitch angle θ, and yaw angle ψ remain bounded within $(-90°, 90°)$. Consequently, the Jacobian matrix $J(z)$ maintains positive definiteness. These angular limitations originate from engineering practices and design requirements ensuring UUV operational effectiveness across diverse tasks and environments. Hence, Assumption 9.2 remains valid. Assumption 9.3 represents a standard assumption for handling unknown dead-zone inputs, establishing the bounded nature of d in (9.4).

Remark 9.2 In this section, $g(z^{(0\sim1)}) + g^T(z^{(0\sim1)}) = m(t)(J(z)M^{-1} + (M^{-1})^T(J(z))^T)$ is uniformly positive definite and the controllability condition holds for the FAUUVs system. Consequently, the norm of the unknown positive definite matrix $g(z^{(0\sim1)})$ possesses a positive lower bound. Moreover, the external time-varying disturbances τ_d remain bounded (Liu et al. 2022b). By combining Remark 9.1 with the analysis, it follows that τ_μ is bounded and satisfies $\|\tau_\mu\| \leq \bar{\tau}_\mu$, where $\bar{\tau}_\mu$ is an unknown positive constant.

Lemma 9.1 (Zhao et al (2022a)) *For any positive constant c, the following function:*

$$I(x) = \frac{\sqrt{c}x}{\sqrt{1 - x^2}},$$

is strictly monotonically increasing for any scalar variable x over the interval $(-1, 1)$.

Based on Lemma 9.1, we construct the following form of prescribed function:

$$I_i\left(\frac{1}{\alpha_i(t)}\right) = \frac{\sqrt{c_i}}{\sqrt{(\alpha_i(t))^2 - 1}}, \tag{9.7}$$

where $c_i, i = 1, 2, \ldots, 6$ are positive constants, $\alpha_i(t) : [0, \infty), i = 1, 2, \ldots, 6$ are $\mathscr{C}^2$ monotonically increasing function and $\dot{\alpha}_i(t), \ddot{\alpha}_i(t)$ are bounded and piecewise continuous functions. Besides, $\alpha_i(0) = 1$ and $\lim_{t \to \infty} \alpha_i(t) = \frac{1}{a_{fi}}$ with the constant $0 < a_{fi} \ll 1$. Therefore, $I_i(\frac{1}{\alpha_i(t)})$ is a monotonically decreasing function and satisfies the following properties:

$$I_i\left(\frac{1}{\alpha_i(t)}\right) = \begin{cases} \infty, & t = 0, \\ \frac{\sqrt{c_i a_{fi}}}{\sqrt{1-a_{fi}^2}}, & t \to \infty. \end{cases} \tag{9.8}$$

According to the above analysis, if the tracking error $e_i = z_i - \eta_{ri}, i = 1, 2, \ldots, 6$ is confined within the prescribed region $I(-\frac{1}{\alpha_i(t)}) < e_i < I(\frac{1}{\alpha_i(t)})$, the control objective is achieved. Therefore, in this chapter, to guarantee that the controller to be developed is independent of initial conditions, we propose the following error transformation:

$$\Upsilon_i(e_i) = \frac{e_i}{\sqrt{e_i^2 + c_i}}. \tag{9.9}$$

It is seen from the above equation that $\Upsilon_i(e_i)$ is a strictly monotonical function and has the properties:

$$\begin{cases} \Upsilon_i(e_i) \to -1, & e_i \to -\infty, \\ \Upsilon_i(e_i) = 0, & e_i = 0, \\ \Upsilon_i(e_i) \to 1, & e_i \to \infty. \end{cases} \tag{9.10}$$

Upon utilizing the definitions of $\alpha_i(t)$ and $\Upsilon_i(e_i)$, the following barrier function is introduced:

$$s_i(t) = \frac{\delta_i(t)}{1 - \delta_i^2(t)}, \tag{9.11}$$

with $\delta_i(t) = \alpha_i(t) \Upsilon_i(e_i)$.

Remark 9.3 According to the expression of $s_i(t)$, we know that

$$s_i(t) \to \begin{cases} +\infty, & \delta_i(t) \to 1, \\ -\infty, & \delta_i(t) \to -1. \end{cases}$$

Furthermore, it is also noted that $\delta_i(0) = \alpha_i(0)\Upsilon_i(e_i(0)) = \Upsilon(e_i(0)) \in (-1, 1)$. The key implication of this property regarding $s_i(t)$ is that for any bounded initial tracking error $e_i(0)$, if $s_i(t)$ remains bounded, then $|\delta_i(t)| < 1$ for $t \geq 0$ is naturally maintained.

Remark 9.4 The preceding analysis demonstrates that $\Upsilon_i(e)$ satisfies $-\frac{1}{\alpha_i(t)} < \Upsilon_i(e_i) < \frac{1}{\alpha_i(t)}$. From Lemma 9.1, it follows that $I[\Upsilon_i(e_i)] = e_i$. Consequently, $I_i(-\frac{1}{\alpha_i(t)}) < I_i[\Upsilon_i(e_i)] = e_i < I_i(\frac{1}{\alpha_i(t)})$, and the challenge of maintaining the tracking error within the prescribed region simplifies to ensuring $s_i(t)$ remains bounded for $t \geq 0$.

9.3 Adaptive Tracking Error-Constraint Controller Design

This section develops a full-actuation framework-based adaptive control scheme to fulfill the prescribed control requirements. The closed-loop system's asymptotic stability is rigorously established via Lyapunov-based stability analysis.

The principal conclusions are derived through temporal differentiation of the sliding surface variable $s_i(t)$. As a preliminary step, consider:

$$\dot{s}_i(t) = \beta_i(\delta_i(t))\dot{\delta}_i(t), \tag{9.12}$$

with $\beta_i(\delta_i(t)) = \frac{1+\delta_i(t)^2}{(1-\delta_i(t)^2)^2}$. Noting that the derivative of $\delta_i(t)$ as given is

$$\dot{\delta}_i(t) = \dot{\alpha}_i\Upsilon_i + \alpha_i\dot{\Upsilon}_i, \tag{9.13}$$

and

$$\dot{\Upsilon}_i(e_i) = \rho_i(e_i)\dot{e}_i, \tag{9.14}$$

where the nonlinear function $c_i(e_i) = \frac{c_i}{(e_i^2+c_i)^{\frac{3}{2}}}$ serves as an analytically tractable component for control synthesis.

Substituting (9.13) into (9.12), we obtain

$$\dot{s}_i(t) = \beta_i(\dot{\alpha}_i\Upsilon_i + \alpha_i\rho_i\dot{e}_i) = \beta_i\dot{\alpha}_i\Upsilon_i + \beta_i\alpha_i\rho_i\dot{e}_i = \mu_{1i} + \mu_{2i}\dot{e}_i, \tag{9.15}$$

with $\mu_{1i} = \beta_i\dot{\alpha}_i\Upsilon_i$, $\mu_{2i} = \beta_i\alpha_i\rho_i$. Then,

$$\ddot{s}(t) = \begin{bmatrix} \dot{\mu}_{11} \\ \dot{\mu}_{12} \\ \vdots \\ \dot{\mu}_{16} \end{bmatrix} + \begin{bmatrix} \dot{\mu}_{21} & & \\ & \dot{\mu}_{22} & \\ & & \ddots \\ & & & \dot{\mu}_{26} \end{bmatrix} \begin{bmatrix} \dot{e}_1 \\ \dot{e}_2 \\ \vdots \\ \dot{e}_6 \end{bmatrix} + \begin{bmatrix} \mu_{21} & & \\ & \mu_{22} & \\ & & \ddots \\ & & & \mu_{26} \end{bmatrix} \begin{bmatrix} \ddot{e}_1 \\ \ddot{e}_2 \\ \vdots \\ \ddot{e}_6 \end{bmatrix}$$

$$= \dot{\mu}_1 + \dot{\mu}_2\dot{e} + \mu_2\ddot{e} = F\left(e^{(0\sim1)}\right) + \mu_2\ddot{e}, \tag{9.16}$$

with $s = [s_1, s_2, \ldots, s_6]^T$ and $F\left(e^{(0\sim1)}\right) = \dot{\mu}_1 + \dot{\mu}_2\dot{e}$.

Combined with (9.6) and (9.16), the ECFAS of UUVs can now be expressed as

$$\ddot{s} = F\left(e^{(0\sim1)}\right) + \mu_2\left(f\left(z^{(0\sim1)}\right) + F + g\left(z^{(0\sim1)}\right)\tau + \tau_\mu - \ddot{\eta}_r\right). \quad (9.17)$$

The intermediate control law v can be designed as

$$v = \frac{1}{\mu_2}\left(A^{0\sim1}s^{(0\sim1)} + F\left(e^{(0\sim1)}\right)\right) + f\left(z^{(0\sim1)}\right) - \ddot{\eta}_r + v^*, \quad (9.18)$$

which gives the following closed-loop system

$$\ddot{s} + A^{0\sim1}s^{(0\sim1)} = \mu_2\left(F + g\left(z^{(0\sim1)}\right)\tau + \tau_\mu + v - v^*\right), \quad (9.19)$$

where v^* is designed in the following procedure.

Then, we rewrite the above ECFAS (9.17) in the following state-space form:

$$\dot{s}^{(0\sim1)} = \Phi\left(A^{0\sim1}\right)s^{(0\sim1)} + \left[\begin{matrix} 0 \\ \mu_2\left(F + g\left(z^{(0\sim1)}\right)\tau + \tau_\mu + v - v^*\right) \end{matrix}\right]. \quad (9.20)$$

Select Lyapunov function candidate as follows:

$$V_1 = \left(s^{(0\sim1)}\right)^T P\left(s^{(0\sim1)}\right), \quad (9.21)$$

where $P \in \mathbb{R}^{12\times12}$ is a positive definite matrix.

Based on the Lemma 2.5, the time derivative of V_1 is computed as

$$\dot{V}_1 = \left(\dot{s}^{(0\sim1)}\right)^T P\left(s^{(0\sim1)}\right) + \left(s^{(0\sim1)}\right)^T P\left(\dot{s}^{(0\sim1)}\right)$$

$$\leq -\xi\left(s^{(0\sim1)}\right)^T P\left(s^{(0\sim1)}\right) + 2\left(s^{(0\sim1)}\right)^T P_L\mu_2(\Delta\left(z^{(0\sim1)}\right)$$

$$+ g\left(z^{(0\sim1)}\right)\tau + \tau_\mu + v - v^*), \quad (9.22)$$

where $\xi > 0$, $P_L = PL$, $L = [0, \dots, I_r]^T$.

Based on the RBF neural network, the lumped uncertainty term can be parametrically decomposed as $F = W^T\Delta(z^{(0\sim1)}) + \chi(z^{(0\sim1)})$, where $W = [W_1, W_2, \dots, W_6] \in \mathbb{R}^{l\times6}$ denotes the ideal radial basis function neural network weight matrix with l hidden nodes, the vector $\Delta = [\Delta_1, \Delta_2, \dots, \Delta_l]^T \in \mathbb{R}^l$ comprises the activation function values, each Δ_i denotes the Gaussian function with $i = 1, 2, \dots, l$, the term $\chi(z^{(0\sim1)}) \in \mathbb{R}^m$ reflects the approximation error and bounded by $\bar{\chi}$. Consequently, the system dynamics in (9.22) admit the reformulation:

$$\dot{V}_1 \leq -\xi V_1 + 2\left(s^{(0\sim1)}\right)^T P_L \mu_2 \left(W^T \Delta(z^{(0\sim1)})\right.$$
$$\left. + \chi(z^{(0\sim1)}) + g\left(z^{(0\sim1)}\right)\tau + \tau_\mu + \nu - \nu^*\right). \tag{9.23}$$

Let $\bar{\chi} \geq \sup \left|\chi(z^{(0\sim1)})\right|$ define the uncertainty bound. Application of Lemma 2.8 yields the fundamental inequality:

$$2\left(s^{(0\sim1)}\right)^T P_L \mu_2 \left(\chi(z^{(0\sim1)}) + \tau_\mu\right)$$
$$\leq \frac{2\Psi \left(s^{(0\sim1)}\right)^T P_L \mu_2 \mu_2^T P_L^T s^{(0\sim1)}}{\sqrt{\left(\left(s^{(0\sim1)}\right)^T P_L \mu_2 \mu_2^T P_L^T s^{(0\sim1)}\right) + \varepsilon^2}} + 2\varepsilon\Psi, \tag{9.24}$$

where the composite uncertainty bound is characterized by $\Psi := \bar{\chi} + \bar{\tau}_\mu$, with ε representing a positive uniformly continuous bounded scalar function. This parameterization yields:

$$\dot{V}_1 \leq -\xi V_1 + 2\left(s^{(0\sim1)}\right)^T P_L \mu_2 \left(W^T \Delta(z^{(0\sim1)}) + g\left(z^{(0\sim1)}\right)\tau + \nu - \nu^*\right)$$
$$+ \frac{2\Psi \left(s^{(0\sim1)}\right)^T P_L \mu_2 \mu_2^T P_L^T s^{(0\sim1)}}{\sqrt{\left(\left(s^{(0\sim1)}\right)^T P_L \mu_2 \mu_2^T P_L^T s^{(0\sim1)}\right) + \varepsilon^2}} + 2\varepsilon\Psi. \tag{9.25}$$

To compensate for parametric uncertainties and exogenous disturbances, the composite control law ν^* is synthesized through adaptive robust design as:

$$\nu^* = \hat{W}^T \Delta(z^{(0\sim1)}) + \frac{\hat{\Psi} \mu_2^T P_L^T s^{(0\sim1)}}{\sqrt{\left(\left(s^{(0\sim1)}\right)^T P_L \mu_2 \mu_2^T P_L^T s^{(0\sim1)}\right) + \varepsilon^2}}, \tag{9.26}$$

where $\hat{\Psi} = \Psi - \tilde{\Psi}$ is the estimation value of Ψ and $\tilde{\Psi}$ is the estimation error. The actual control law is chosen as

$$\tau = -\frac{\hat{\Theta}^2 \nu \nu^T \mu_2^T P_L^T s^{(0\sim1)}}{\sqrt{\left(\hat{\Theta}\left(s^{(0\sim1)}\right)^T P_L \mu_2 \nu\right)^2 + \varepsilon^2}}. \tag{9.27}$$

From (9.25) and the actual controller (9.27), it can be derived that

$$
2\left(s^{(0\sim1)}\right)^T P_L \mu_2 g\left(z^{(0\sim1)}\right)\tau
$$

$$
= -\frac{2\hat{\Theta}^2 \left\|g\left(z^{(0\sim1)}\right)\right\| \left(s^{(0\sim1)}\right)^T P_L \mu_2 v v^T \mu_2^T P_L^T s^{(0\sim1)}}{\sqrt{\left(\hat{\Theta}\left(s^{(0\sim1)}\right)^T P_L \mu_2 v\right)^2 + \varepsilon^2}}
$$

$$
\leq -\frac{2\underline{g}\left(\hat{\Theta}\left(s^{(0\sim1)}\right)^T P_L \mu_2 v\right)^2}{\sqrt{\left(\hat{\Theta}\left(s^{(0\sim1)}\right)^T P_L \mu_2 v\right)^2 + \varepsilon^2}}
$$

$$
\leq 2\underline{g}\varepsilon - 2\underline{g}\hat{\Theta}\left(s^{(0\sim1)}\right)^T P_L \mu_2 v, \tag{9.28}
$$

where $\hat{\Theta} = \Theta - \tilde{\Theta}$ is the estimation value of $\Theta = \dfrac{1}{g}$, $\tilde{\Theta}$ is the estimation error, and $\underline{g} > 0$ is the lower bound of $\left\|g(z^{(0\sim1)})\right\|$. Then, we can get

$$
\dot{V}_1 \leq -\xi V_1 + 2\left(s^{(0\sim1)}\right)^T P_L \mu_2 \widetilde{W}^T \Delta(z^{(0\sim1)}) + 2\underline{g}\varepsilon - 2\underline{g}\hat{\Theta}\left(s^{(0\sim1)}\right)^T P_L \mu_2 v
$$

$$
- \frac{2\widehat{\Psi}\left(s^{(0\sim1)}\right)^T P_L \mu_2 \mu_2^T P_L^T s^{(0\sim1)}}{\sqrt{\left(\left(s^{(0\sim1)}\right)^T P_L \mu_2 \mu_2^T P_L^T s^{(0\sim1)}\right) + \varepsilon^2}} + 2\left(s^{(0\sim1)}\right)^T P_L \mu_2 v
$$

$$
+ \frac{2\Psi\left(s^{(0\sim1)}\right)^T P_L \mu_2 \mu_2^T P_L^T s^{(0\sim1)}}{\sqrt{\left(\left(s^{(0\sim1)}\right)^T P_L \mu_2 \mu_2^T P_L^T s^{(0\sim1)}\right) + \varepsilon^2}} + 2\varepsilon\Psi
$$

$$
\leq -\xi V_1 + 2\left(s^{(0\sim1)}\right)^T P_L \mu_2 \widetilde{W}^T \Delta(z^{(0\sim1)}) + 2\underline{g}\varepsilon - 2\underline{g}\hat{\Theta}\left(s^{(0\sim1)}\right)^T P_L \mu_2 v
$$

$$
+ \frac{2\widetilde{\Psi}\left(s^{(0\sim1)}\right)^T P_L \mu_2 \mu_2^T P_L^T s^{(0\sim1)}}{\sqrt{\left(\left(s^{(0\sim1)}\right)^T P_L \mu_2 \mu_2^T P_L^T s^{(0\sim1)}\right) + \varepsilon^2}} + 2\varepsilon\Psi + 2\left(s^{(0\sim1)}\right)^T P_L \mu_2 v.
$$

$$
\tag{9.29}
$$

Remark 9.5 The controller developed in this section employs a two-stage methodology. Initially, the intermediate controller v in (9.18) is derived via the ECFAS approach. Subsequently, to counteract model uncertainties, external disturbances, and unknown dead-zone inputs, the final control signal (9.27) is formulated. Through subsequent analysis using the framework established herein, the UUVs' nonlinearities are effectively mitigated, achieving asymptotic stability.

9.4 Stability Analysis with Error-Constraint

The closed-loop stability theorem is formally established through the proposed control synthesis framework, with comprehensive stability analysis conducted for FAUUVs under the developed control architecture.

Theorem 9.1 *Consider the fully actuated unmanned underwater vehicle systems consisting of model uncertainties, external disturbances and unknown dead-zone inputs. Under Assumptions 9.1–9.4, the feasible virtual control signal (9.18) and (9.26), the actual controller (9.27) and the following adaptive laws:*

$$\dot{\hat{\Theta}} = 2 \left(s^{(0 \sim 1)} \right)^T P_L \mu_2 v, \tag{9.30}$$

$$\dot{\hat{\Psi}} = \frac{2 \left(s^{(0 \sim 1)} \right)^T P_L \mu_2 \mu_2^T P_L^T s^{(0 \sim 1)}}{\sqrt{\left(\left(s^{(0 \sim 1)} \right)^T P_L \mu_2 \mu_2^T P_L^T s^{(0 \sim 1)} \right) + \varepsilon^2}}, \tag{9.31}$$

and

$$\dot{\hat{W}}_i = 2 \Gamma_i \Delta(z^{(0 \sim 1)}) \left(s^{(0 \sim 1)} \right)^T P_{Li} \mu_{2i}, \tag{9.32}$$

with $P_L = [P_{L1}, P_{L2}, \dots, P_{L6}] \in \mathbb{R}^{12 \times 6}$ and $P_{Li} \in \mathbb{R}^{12}, i = 1, 2, \dots, 6$ and $\Gamma_i \in \mathbb{R}^{l \times l}$ is a symmetric positive definite matrix. Furthermore, all the tracking error signal $e_i, i = 1, 2, \dots, 6$ converge to zero asymptotically.

Proof To analyze the entire adaptive controller, employ the Lyapunov function $V = V_1 + \frac{1}{2} \sum_{i=1}^{6} \tilde{W}_i^T \Gamma_i^{-1} \tilde{W}_i + \frac{1}{2} \underline{g} \tilde{\Theta}^2 + \frac{1}{2} \tilde{\Psi}^2$, where $\tilde{W}_i = W_i - \hat{W}_i \in \mathbb{R}^l, i = 1, 2, \dots, 6$ is the ith approximation weight error vector.

From (9.29), the time derivative of Lyapunov function V is shown as

$$\dot{V} \leq -\xi V_1 + 2 \left(s^{(0 \sim 1)} \right)^T P_L \mu_2 \tilde{W}^T \Delta(z^{(0 \sim 1)}) + 2 \underline{g} \varepsilon - 2 \underline{g} \hat{\Theta} \left(s^{(0 \sim 1)} \right)^T P_L \mu_2 v$$

$$+ \frac{2 \tilde{\Psi} \left(s^{(0 \sim 1)} \right)^T P_L \mu_2 \mu_2^T P_L^T s^{(0 \sim 1)}}{\sqrt{\left(\left(s^{(0 \sim 1)} \right)^T P_L \mu_2 \mu_2^T P_L^T s^{(0 \sim 1)} \right) + \varepsilon^2}} + 2 \varepsilon \Psi$$

$$- \sum_{i=1}^{6} \tilde{W}_i^T \Gamma_i^{-1} \dot{\hat{W}}_i - \underline{g} \tilde{\Theta} \dot{\hat{\Theta}} - \tilde{\Psi} \dot{\hat{\Psi}} + 2 \left(s^{(0 \sim 1)} \right)^T P_L \mu_2 v. \tag{9.33}$$

By substituting the adaptive network updating laws (9.30), (9.31) and (9.32) for (9.33), we can obtain

$$\dot{V} \leq -\xi V_1 + 2\underline{g}\varepsilon + 2\varepsilon\Psi. \tag{9.34}$$

Integrating the function $\dot{V}$ over the interval $[t_0, t]$, we can obtain

$$0 \leq V(t) \leq V(t_0) - \xi \int_{t_0}^{t} V_1(t_1)dt_1 + 2\left(\underline{g} + \Psi\right)\bar{\varepsilon}, \tag{9.35}$$

which means that $s, \dot{s}, \tilde{\Theta}, \tilde{\Psi}$ and $\tilde{W}_i \in \mathbb{R}^l, i = 1, 2, \ldots, 6$ are bounded. Furthermore, (9.35) can be derived as

$$\xi \int_{t_0}^{t} \left(s^{(0\sim1)}\right)^T (t_1)P\left(s^{(0\sim1)}\right)(t_1)dt_1 \leq V(t_0) + 2\left(\underline{g} + \Psi\right)\bar{\varepsilon}. \tag{9.36}$$

Application of Barbalat's lemma rigorously establishes $\lim_{t\to\infty} s_i(t) = 0$, guaranteeing asymptotic convergence of UUV trajectories to reference targets. This completes the closed-loop stability demonstration.

Remark 9.6 In practical control applications for UUVs, the algorithm introduced in this section allows for the flexible selection of $\varepsilon = e^{-ct}$, where c represents a positive constant. As time t approaches infinity, the resulting decrease in ε approaches zero. Thus, the output tracking error asymptotically converges to zero, effectively reducing the influence of model uncertainties, external disturbances, and unknown dead-zone inputs.

Remark 9.7 In this section, a unique adaptive control strategy incorporating the ECFAS model is presented. Based on Lyapunov stability theory analysis, it is established that the system's output tracking error asymptotically converges to zero. This confirms the proposed approach's effectiveness in countering the effects of model uncertainties, external disturbances, and unknown dead-zone inputs. To

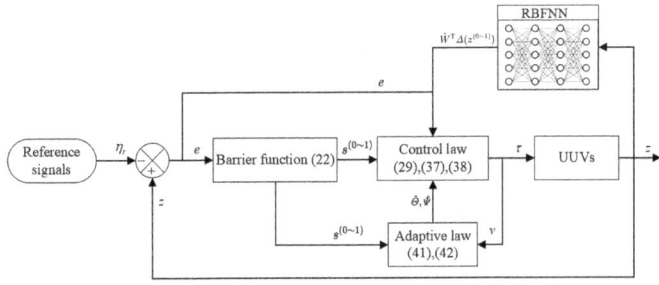

Fig. 9.3 The schematic diagram of the proposed approach

exemplify the practical applicability of Theorem 9.1, Fig. 9.3 provides a schematic representation of the proposed methodology. The next section then validates the efficacy of the ECFAS approach through simulation studies.

9.5   Simulation Example of Adaptive Tracking Control Performance

The closed-loop control performance is experimentally evaluated for the UUV system (9.1) through reference trajectory tracking. The desired trajectory is defined as $\eta_r = [\mathfrak{s}(0.1t)+1, \mathfrak{c}(0.1t)+1, 0.2t+1, 0, 0, 0]^T$. The hydrodynamic environment features a time-varying current field $\mu_c = [0.1\mathfrak{s}(t), 0.1\mathfrak{c}(t), 0, 0, 0, 0]^T$, with additional environmental disturbances modeled as:

$$\tau_d = [0.5\mathfrak{s}(0.3t), 0.08\mathfrak{c}(0.2t), 0.9\,|\mathfrak{c}(0.1t)|, 0, 0, 0]^T. \tag{9.37}$$

To account for uncertainties such as unknown hydrodynamic parameters, modeling errors, parameter perturbation, and unknown dead-zone inputs in the UUVs, the following model uncertainties are considered:

Firstly, the damping matrix $D(\mu)$ incorporates the effect of unknown hydrodynamic parameters. It is formulated as:

$$D(\mu) = \begin{bmatrix} u\,(2\,|0.1\mathfrak{s}(t) - u| + 0.2) \\ v\,(5\,|0.1\mathfrak{c}(t) - v| + 0.4) \\ \omega\,(6\,|\omega| + 2) \\ p\,(8\,|p| + 0.3) \\ q\,(|q| + 0.5) \\ r\,(|r| + 0.4) \end{bmatrix}. \tag{9.38}$$

Considering the small influence of the additional Coriolis force, the matrix $C_d(\mu)$ is given by:

$C_d(\mu)$

$$= \begin{bmatrix} 0 & 0 & 0 & 0 & 3\omega & 0.3\mathfrak{c}(t) - 3v \\ 0 & 0 & 0 & -3\omega & 0 & 2u - 0.2\mathfrak{s}(t) \\ 0 & 0 & 0 & 3v - 0.3\mathfrak{c}(t) & 0.2\mathfrak{s}(t) - 2u & 0 \\ 0 & 3\omega & 0.3\mathfrak{c}(t) - 3v & 0 & r & -q \\ -3\omega & 0 & 2u - 0.2\mathfrak{s}(t) & -r & 0 & p \\ 3v - 0.3\mathfrak{c}(t) & 0.2\mathfrak{s}(t) - 2u & 0 & q & -p & 0 \end{bmatrix}. \tag{9.39}$$

Table 9.1 Parameters of the UUVs

Parameter	Symbols	Value	Unit
Mass	m_0	51	kg
Center radius	$[r_x, r_y, r_z]$	$[0.37, 0.24, 0.15]$	m
Center of gravity coordinates	$[x_G, y_G, z_G]$	$[0, 0, 0.15]$	m
Center of buoyancy Coordinates	$[x_B, y_B, z_B]$	$[0, 0, -0.12]$	m
Density of water	ρ_0	1000	kg/m^3
Volume of UUV	V_0	0.49	m^3
Gravitational acceleration	g	9.98	m/s^2

The dead-zone input characteristic $D_z(\tau_i)$ is defined as follows:

$$D_z(\tau_i) = \begin{cases} 5(\tau_i - 0.2), & \tau \geq 0.2, \\ 0, & -0.5 < \tau_i < 0.2, \\ 4(\tau_i + 0.5), & \tau \leq -0.5. \end{cases} \tag{9.40}$$

The dynamic model described by Eq. (9.1) incorporates these essential parameters, where the mathematical formulation considers both inertial and environmental interaction forces. The parameter values listed in Table 9.1 were obtained through experimental validation and computational fluid dynamics simulations.

The initial relative position, relative attitude, relative velocity, relative angular velocity of the UUVs are set, respectively, as

$$\eta(0) = [-0.3, -0.2, 0.1, -0.1, 0.2, 0.2]^T, \mu(0) = [0.1, 0.1, 0.1, 0.1, 0.1, 0.1]^T.$$

The RBFNN employs a hidden layer with 5×6 nodes. The Gaussian basis function width is configured as $\epsilon_i = 1$ for $i = 1, 2, \ldots, 6$, with receptive field centers $\kappa_i \in \mathbb{R}^j$ distributed over $[-8, 8]$. Initial weights are assigned as $\hat{W}_i(0) = 1$. This configuration establishes the RBFNN architecture and initializes its parameters.

Following the operational constraints of UUVs, the time-varying scaling functions are formulated as:

$$\alpha_1 = \frac{20}{17e^{-0.25t} + 3}, \alpha_2 = \frac{20}{17e^{-0.3t} + 3}, \alpha_3 = \frac{20}{17e^{-0.4t} + 3},$$

$$\alpha_4 = \frac{20}{17e^{-0.6t} + 3}, \alpha_5 = \frac{20}{17e^{-0.4t} + 3}, \alpha_6 = \frac{20}{17e^{-0.6t} + 3}.$$

In the controller design process, the coefficient vector $A^{0 \sim 1}$ is selected as:

$$A = \begin{bmatrix} 10I_{6\times6} & 5I_{6\times6} \end{bmatrix}, \tag{9.41}$$

Fig. 9.4 The UUVs tracks the reference signals asymptotically. (**a**) Tracking track of the positions. (**b**) Target position tracking in 3-D space

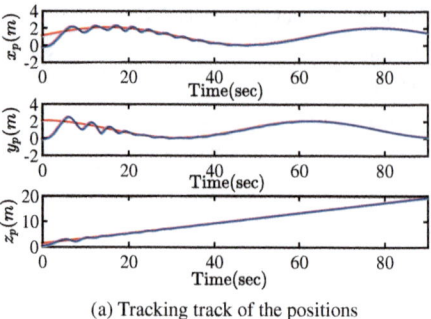

(a) Tracking track of the positions

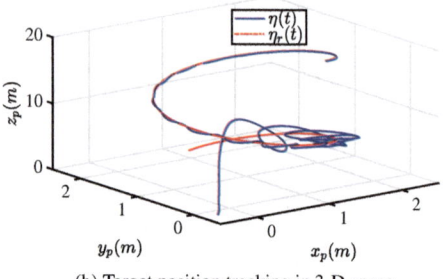

(b) Target position tracking in 3-D space

where $I_{6\times6}$ represents a 6×6 identity matrix. The matrix $\Phi(A)$ is then defined as:

$$\Phi\,(A) = \begin{bmatrix} 0_{6\times6} & I_{6\times6} \\ -10I_{6\times6} & -5I_{6\times6} \end{bmatrix}. \tag{9.42}$$

By integrating these expressions, we construct time-varying scaling functions and specify the coefficient vectors required for controller design.

Consider the Lyapunov function in (9.21), we choose $\xi = 1$ and obtain

$$P = \begin{bmatrix} 4.5998I_{6\times6} & 0.0660I_{6\times6} \\ 0.0660I_{6\times6} & 0.0735I_{6\times6} \end{bmatrix}. \tag{9.43}$$

Numerical simulations of the proposed controller (Theorem 9.1) are demonstrated in Figs. 9.4, 9.5, 9.6, 9.7, 9.8, and 9.9. Figure 9.4 illustrates the relative position response, confirming asymptotic reference trajectory tracking for UUVs. The constrained position and attitude tracking errors in Figs. 9.5 and 9.6 reveal that all errors remain strictly within prescribed boundaries, satisfying both controller specifications and operational reliability requirements. Actuator-generated control signals are presented in Figs. 9.7 and 9.8, where τ_1 to τ_3 denote positional control forces and τ_4 to τ_6 represent attitude control torques. Figure 9.9 validates the boundedness of adaptive laws $\widehat{\Theta}(t)$ and $\widehat{\Psi}(t)$. The simulation results demonstrate that the proposed adaptive error-constrained tracking controller effectively restricts

Fig. 9.5 The tracking error
of x_p, y_p, z_p under
constrained conditions. (**a**)
The tracking error of x_p. (**b**)
The tracking error of y_p. (**c**)
The tracking error of z_p

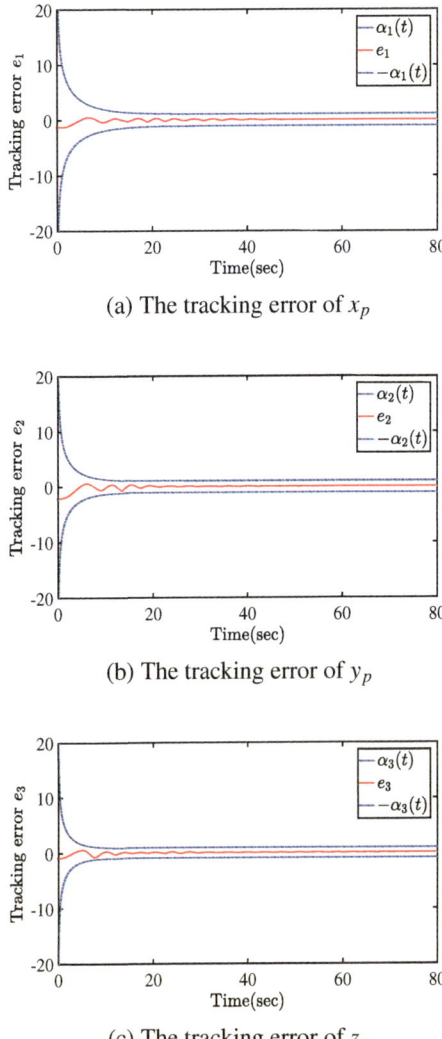

(a) The tracking error of x_p

(b) The tracking error of y_p

(c) The tracking error of z_p

the tracking error e of unmanned underwater vehicles (UUVs), while maintaining transient and steady-state precision. The controller exhibits adaptive compensation capabilities against unknown hydrodynamic parameters, modeling inaccuracies, input dead-zones, and parametric perturbations.

Fig. 9.6 The tracking error of φ, θ, ψ under constrained conditions. (**a**) The tracking error of φ. (**b**) The tracking error of θ. (**c**) The tracking error of ψ

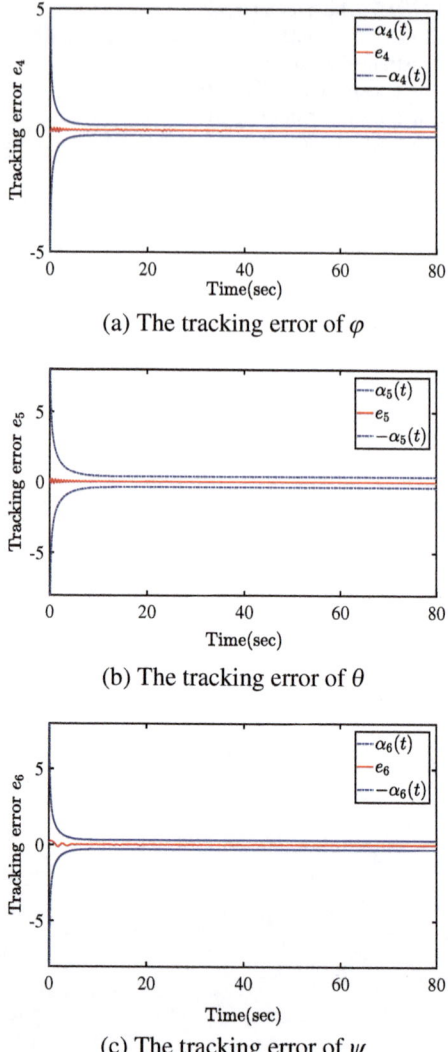

(a) The tracking error of φ

(b) The tracking error of θ

(c) The tracking error of ψ

Fig. 9.7 The control vector
x, y, z. (**a**) The force in the
x-direction. (**b**) The force in
the y-direction. (**c**) The force
in the z-direction

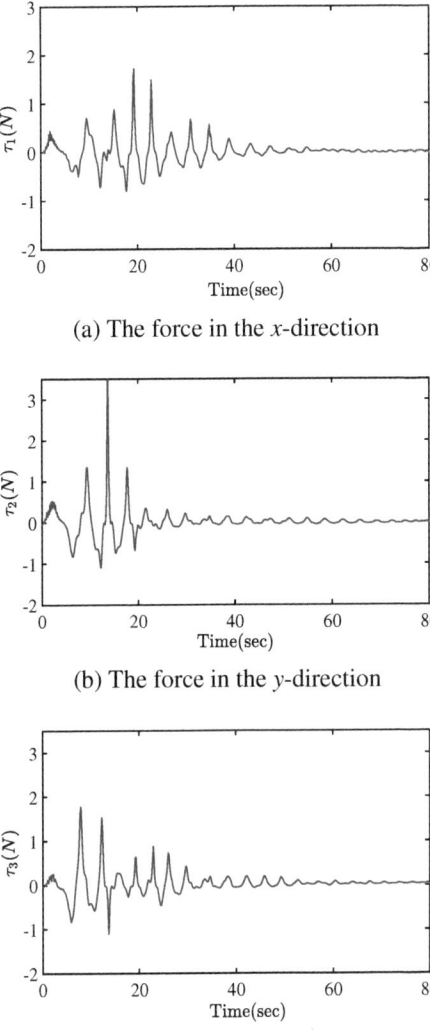

(a) The force in the x-direction

(b) The force in the y-direction

(c) The force in the z-direction

Fig. 9.8 The control vector x, y, z. (**a**) The torque on the x-axis. (**b**) The torque on the y-axis. (**c**) The torque on the z-axis

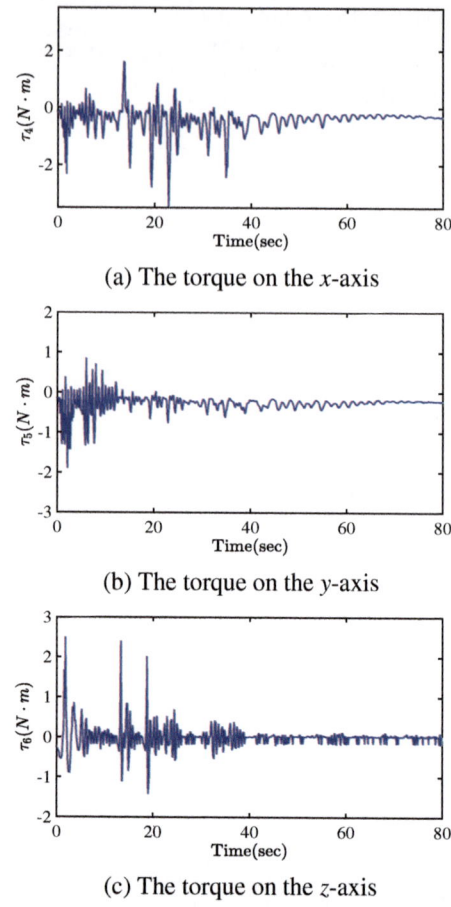

(a) The torque on the x-axis

(b) The torque on the y-axis

(c) The torque on the z-axis

Fig. 9.9 The adaptive estimation. (**a**) The response of $\widehat{\Theta}(t)$. (**b**) The response of $\widehat{\Psi}(t)$

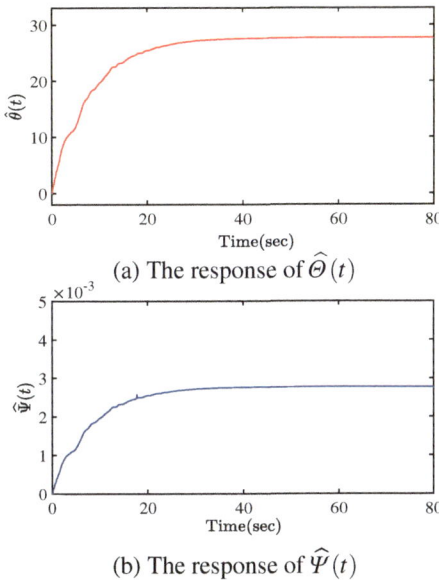

(a) The response of $\widehat{\Theta}(t)$

(b) The response of $\widehat{\Psi}(t)$

9.6   Notes and References

In this chapter, a novel adaptive position tracking error constrained controller for UUVs is established based on ECFAS approach, the constant matrices $A^{0\sim1}$ are obtained to convert the closed-loop system to a linear constant system with a desired eigenstructure, which saves complicated calculations caused by recursive virtual controller design. A special yet unique prescribed function is applied by employing the tracking error dependent normalized function and barrier function along with time-varying scaling, which is independent of the initial conditions of nonlinear systems. Since the accurate model of the UUVs often cannot be obtained, we propose an adaptive algorithm based on NNs to approximate the model uncertainty. Furthermore, a positive time-varying integral function is introduced to completely eliminate the effect of the residual effect caused by model uncertainties and unknown dead-zone inputs. Through the above control strategy, the position tracking error constraints can also be strictly maintained, and the UUVs can track the trajectory asymptotically, which is of value in practical application. The final numerical simulation demonstrates the effectiveness of the proposed controller.

References

Adıgüzel, F., & Yalçın, Y. (2022). Immersion and invariance disturbance observer-based nonlinear discrete-time control for fully actuated mechanical systems. *International Journal of Systems Science, 53*(2), 388–401.

Agarwal, M., & Seborg, D. E. (1987). Self-tuning controllers for nonlinear systems. *Automatica, 23*(2), 209–214.

An, H., Xia, H., & Wang, C. (2017). Barrier Lyapunov function-based adaptive control for hypersonic flight vehicles switched systems. *Nonlinear Dynmaic, 88*, 1833–1853.

Andrieu, V., Praly, L., & Astolfi, A. (2008). Homogeneous approximation, recursive observer design, and output feedback. *SIAM Journal on Control and Optimization, 47*(4), 1814–1850.

Åström, K. J., & Wittenmark, B. (1973). On self tuning regulators. *Automatica, 9*(2), 185–199.

Bechlioulis, C. P., & Rovithakis, G. A. (2008a). Prescribed performance adaptive control of siso feedback linearizable systems with disturbances. In *2008 16th Mediterranean Conference on Control and Automation* (pp. 1035–1040)

Bechlioulis, C. P., & Rovithakis, G. A. (2008b). Robust adaptive control of feedback linearizable MIMO nonlinear systems with prescribed performance. *IEEE Transactions on Automatic Control, 53*(9), 2090–2099.

Bekiaris-Liberis, N., & Krstic, M. (2010). Stabilization of linear strict-feedback systems with delayed integrators. *Automatica, 46*(11), 1902–1910.

Bekiaris-Liberis, N., & Krstic, M. (2012). Compensation of state-dependent input delay for nonlinear systems. *IEEE Transactions on Automatic Control, 58*(2), 275–289.

Bekiaris-Liberis, N., & Krstic, M. (2016). Stability of predictor-based feedback for nonlinear systems with distributed input delay. *Automatica, 70*, 195–203.

Bhat, S., & Bernstein, D. (1995). Lyapunov analysis of finite-time differential equations. In *Proceedings of 1995 American Control Conference - ACC'95* (vol. 3, pp. 1831–1832).

Bhat, S., & Bernstein, D. (1997). Finite-time stability of homogeneous systems. In *Proceedings of the 1997 American Control Conference (Cat. No.97CH36041)* (vol. 4, pp. 2513–2514)

Bhat, S. P., & Bernstein, D. S. (2000). Finite-time stability of continuous autonomous systems. *SIAM Journal on Control and optimization, 38*(3), 751–766.

Bhat, S. P., & Bernstein, D. S. (2005). *Geometric homogeneity with applications to finite-time stability* (vol. 17, pp. 101–127). Springer.

Bu, X., Wu, X., Zhu, F., Huang, J., Ma, Z., & Zhang, R. (2015). Novel prescribed performance neural control of a flexible air-breathing hypersonic vehicle with unknown initial errors. *ISA Transactions, 59*, 149–159.

Cai, M., He, X., & Zhou, D. (2023a). An active fault tolerance framework for uncertain nonlinear high-order fully-actuated systems. *Automatica, 152*, 110969.

Cai, M., He, X., & Zhou, D. (2023b). Fault-tolerant tracking control for nonlinear observer-extended high-order fully-actuated systems. *Journal of the Franklin Institute, 360*(1), 136–153.

Cai, M., He, X., & Zhou, D. (2023c). Low-power fault-tolerant control for nonideal high-order fully actuated systems. *IEEE Transactions on Systems, Man, and Cybernetics: Systems, 53*(8), 4875–4887.

Cai, M., Shi, P., & Yu, J. (2024). Adaptive neural finite-time control of non-strict feedback nonlinear systems with non-symmetrical dead-zone. *IEEE Transactions on Neural Networks and Learning Systems, 35*(1), 1409–1414.

Cao, B., & Nie, X. (2021). Event-triggered adaptive neural networks control for fractional-order nonstrict-feedback nonlinear systems with unmodeled dynamics and input saturation. *Neural Networks, 142*, 288–302.

Cao, Y., Cao, J., & Song, Y. (2022a). Practical prescribed time control of euler-lagrange systems with partial/full state constraints: a settling time regulator-based approach. *IEEE Transactions on Cybernetics, 52*(12), 13096–13105.

Cao, Y., Cao, J., & Song, Y. (2022b). Practical prescribed time tracking control over infinite time interval involving mismatched uncertainties and non-vanishing disturbances. *Automatica, 136*, 110050.

Chen, J., & Hua, C. (2022). Adaptive full-state-constrained control of nonlinear systems with deferred constraints based on nonbarrier Lyapunov function method. *IEEE Transactions on Cybernetics, 52*(8), 7634–7642.

Chen, M., Liu, X., & Wang, H. (2015). Adaptive robust fault-tolerant control for nonlinear systems with prescribed performance. *Nonlinear Dynamics, 81*(4), 1727–1739.

Chen, M., Wang, H., & Liu, X. (2019). Adaptive fuzzy practical fixed-time tracking control of nonlinear systems. *IEEE Transactions on Fuzzy Systems, 29*(3), 664–673.

Chen, M., Wang, H., & Liu, X. (2021). Adaptive practical fixed-time tracking control with prescribed boundary constraints. *IEEE Transactions on Circuits and Systems I: Regular Papers, 68*(4), 1716–1726.

Chen, S., Wang, W., Fan, J., & Ji, Y. (2023). Impact angle constraint guidance law using fully-actuated system approach. *Aerospace Science and Technology, 136*, 108220.

Cheng, C., Zhang, Y., & Liu, S. (2019). Neural observer-based adaptive prescribed performance control for uncertain nonlinear systems with input saturation. *Neurocomputing, 370*, 94–103.

Clarke, D. W., & Gawthrop, P. J. (1975). Self-tuning controller. In *Proceedings of the Institution of Electrical Engineers* (vol. 122, pp. 929–934). IET.

Cui, K.-X., Duan, G.-R., & Hou, M.-Z. (2024). Discrete-time model reference tracking control for a class of combined spacecraft: A high-order fully actuated system approach. *IEEE Transactions on Automation Science and Engineering, 21*(4), 6966–6977.

Cui, M., & Tong, S. (2023). Event-triggered predefined-time output feedback control for fractional-order nonlinear systems with input saturation. *IEEE Transactions on Fuzzy Systems, 31*(12), 4397–4409.

Cui, R., Yang, C., Li, Y., et al. (2017). Adaptive neural network control of auvs with control input nonlinearities using reinforcement learning. *IEEE Transactions on Systems, Man, and Cybernetics: Systems, 47*(6), 1019–1029.

Cui, Y., Duan, G., Liu, X., et al. (2023a). Adaptive fuzzy fault-tolerant control of high-order nonlinear systems: A fully actuated system approach. *International Journal of Fuzzy Systems, 25*(5), 1895–1906.

Cui, Y., Duan, G., Liu, X., & Zheng, H. (2023b). Adaptive fuzzy fault-tolerant control of high-order nonlinear systems: A fully actuated system approach. *International Journal of Fuzzy Systems, 25*(5), 1895–1906.

Dong, R., Hua, C., Li, K., & Meng, R. (2023). Adaptive fault-tolerant control for high-order fully actuated system with full-state constraints. *Journal of the Franklin Institute, 360*(12), 8062–8074.

Duan, G. (2020a). High-order system approaches: I. Fully-actuated systems and parametric designs. *Acta Automatica Sinica, 46*(7), 1333–1345.

Duan, G. (2020b). High-order system approaches: II. Controllability and fully-actuation. *Acta Automatica Sinica, 46*(8), 1571–1581.

Duan, G. (2020c). High-order system approaches: III. Observability and observer design. *Acta Automatica Sinica, 46*(9), 1885–1895.

Duan, G. (2021a). High-order fully actuated system approaches: Part I. Models and basic procedure. *International Journal of Systems Science, 52*(2), 422–435.

Duan, G. (2021b). High-order fully actuated system approaches: Part II. Generalized strict-feedback systems. *International Journal of Systems Science, 52*(3), 437–454.

Duan, G. (2021c). High-order fully actuated system approaches: Part III. Robust control and high-order backstepping. *International Journal of Systems Science, 52*(5), 952–971.

Duan, G. (2021d). High-order fully actuated system approaches: Part IV. Adaptive control and high-order backstepping. *International Journal of Systems Science, 52*(5), 972–989.

Duan, G. (2021e). High-order fully actuated system approaches: Part V. Robust adaptive control. *International Journal of Systems Science, 52*(10), 2129–2143.

Duan, G. (2021f). High-order fully-actuated system approaches: Part VI. Disturbance attenuation and decoupling. *International Journal of Systems Science, 52*(10), 2161–2181.

Duan, G. (2021g). High-order fully actuated system approaches: Part VII. Controllability, stabilisability and parametric designs. *International Journal of Systems Science, 52*(14), 3091–3114.

Duan, G. (2022a). High-order fully-actuated system approaches: Part IX. Generalised pid control and model reference tracking. *International Journal of Systems Science, 53*(3), 652–674.

Duan, G. (2022b). High-order fully actuated system approaches: Part VIII. Optimal control with application in spacecraft attitude stabilisation. *International Journal of Systems Science, 53*(1), 54–73.

Duan, G. (2022c). High-order fully actuated system approaches: Part X. Basics of discrete-time systems. *International Journal of Systems Science, 53*(4), 810–832.

Duan, G., & Zhao, Q. (2022). Fully actuated system approach for 6DOF spacecraft control based on extended state observer. *Journal of Systems Science and Complexity, 35*(2), 604–622.

Duan, G.-R. (2020d). High-order system approaches: I. Fully-actuated systems and parametric designs. *Acta Automatica Sinica, 46*(7), 1333–1345.

Duan, G.-R. (2024). Fully actuated system approach for control: An overview. *IEEE Transactions on Cybernetics, 54*(12), 7285–7306.

Espitia, N., & Perruquetti, W. (2022). Predictor-feedback prescribed-time stabilization of LTI systems with input delay. *IEEE Transactions on Automatic Control, 67*(6), 2784–2799.

Espitia, N., Steeves, D., Perruquetti, W., & Krstic, M. (2022). Sensor delay-compensated prescribed-time observer for LTI systems. *Automatica, 135*, 110005.

Fan, Y., Li, Y., & Tong, S. (2021). Adaptive finite-time fault-tolerant control for interconnected nonlinear systems. *International Journal of Robust and Nonlinear Control, 31*(5), 1564–1581.

Fridman, E., & Shaked, U. (2002). An improved stabilization method for linear time-delay systems. *IEEE Transactions on Automatic Control, 47*(11), 1931–1937.

Fu, J., Ma, R., & Chai, T. (2015). Global finite-time stabilization of a class of switched nonlinear systems with the powers of positive odd rational numbers. *Automatica, 54*, 360–373.

Fu, J., Ma, R., & Chai, T. (2017). Adaptive finite-time stabilization of a class of uncertain nonlinear systems via logic-based switchings. *IEEE Transactions on Automatic Control, 62*(11), 5998–6003.

Galicki, M. (2015). Finite-time control of robotic manipulators. *Automatica, 51*, 49–54.

Gao, F., Wu, Y., & Zhang, Z. (2019). Global fixed-time stabilization of switched nonlinear systems: A time-varying scaling transformation approach. *IEEE Transactions on Circuits and Systems II: Express Briefs, 66*(11), 1890–1894.

Gao, H., Zhang, T., & Xia, X. (2014). Adaptive neural control of stochastic nonlinear systems with unmodeled dynamics and time-varying state delays. *Journal of the Franklin Institute, 351*(6), 3182–3199.

Gao, T., Liu, Y.-J., Li, D., Tong, S., & Li, T. (2021). Adaptive neural control using tangent time-varying blfs for a class of uncertain stochastic nonlinear systems with full state constraints. *IEEE Transactions on Cybernetics, 51*(4), 1943–1953.

Gao, Y.-F., Sun, X.-M., Wen, C., & Wang, W. (2017). Adaptive tracking control for a class of stochastic uncertain nonlinear systems with input saturation. *IEEE Transactions on Automatic Control, 62*(5), 2498–2504.

Gaudette, D. L., & Miller, D. E. (2014). Stabilizing a SISO LTI plant with gain and delay margins as large as desired. *IEEE Transactions on Automatic Control, 59*(9), 2324–2339.

Guo, C., & Hu, J. (2023a). Fixed-time stabilization of high-order uncertain nonlinear systems: Output feedback control design and settling time analysis. *Journal of Systems Science and Complexity, 36*, 1351–1372.

Guo, C., & Hu, J. (2023b). Time base generator-based practical predefined-time stabilization of high-order systems with unknown disturbance. *IEEE Transactions on Circuits and Systems II: Express Briefs, 70*(7), 2670–2674.

Guo, C., Hu, J., Wu, Y., et al. (2023). Non-singular fixed-time tracking control of uncertain nonlinear pure-feedback systems with practical state constraints. *IEEE Transactions on Circuits and Systems I: Regular Papers, 70*(9), 3746–3758.

Guo, L., Yu, X., Yin, J., & Khoo, S. (2024). A new theorem on finite-time stability of stochastic homogeneous systems and its application. *Asian Journal of Control, 26*(1), 542–548.

He, W., Yin, Z., & Sun, C. (2017). Adaptive neural network control of a marine vessel with constraints using the asymmetric barrier Lyapunov function. *IEEE Transactions on Cybernetics, 47*(7), 1641–1651.

He, X., Li, X., & Song, S. (2023). Prescribed-time stabilization of nonlinear systems via impulsive regulation. *IEEE Transactions on Systems, Man, and Cybernetics: Systems, 53*(2):981–985.

Holloway, J., & Krstic, M. (2019a). Prescribed-time observers for linear systems in observer canonical form. *IEEE Transactions on Automatic Control, 64*(9), 3905–3912.

Holloway, J., & Krstic, M. (2019b). Prescribed-time output feedback for linear systems in controllable canonical form. *Automatica, 107*, 77–85.

Hong, Y. (2002). Finite-time stabilization and stabilizability of a class of controllable systems. *Systems and Control Letters, 46*(4), 231–236.

Hong, Y., Huang, J., Xu, Y. (2001). On an output feedback finite-time stabilization problem. *IEEE Transactions on Automatic Control, 46*(2), 305–309.

Hong, Y., Wang, J., & Cheng, D. (2006). Adaptive finite-time control of nonlinear systems with parametric uncertainty. *IEEE Transactions on Automatic Control, 51*(5), 858–862.

Hu, L., Duan, G., & Hou, M. (2023). Robust adaptive guaranteed cost tracking control for high-order nonlinear systems with uncertainties based on high-order fully actuated system approaches. *International Journal of Robust and Nonlinear Control, 33*(13), 7583–7605.

Hua, C., & Guan, X. (2008). Output feedback stabilization for time-delay nonlinear interconnected systems using neural networks. *IEEE Transactions on Neural Networks, 19*(4), 673–688.

Hua, C., Guan, X., & Shi, P. (2005). Robust backstepping control for a class of time delayed systems. *IEEE Transactions on Automatic Control, 50*(6), 894–899.

Hua, C., Li, K., & Guan, X. (2018). Event-based dynamic output feedback adaptive fuzzy control for stochastic nonlinear systems. *IEEE Transactions on Fuzzy Systems, 26*(5), 3004–3015.

Hua, C., & Li, Y. (2015). Output feedback prescribed performance control for interconnected time-delay systems with unknown prandtl–ishlinskii hysteresis. *Journal of the Franklin Institute, 352*(7), 2750–2764.

Hua, C., Liu, G., Li, L., & Guan, X. (2017a). Adaptive fuzzy prescribed performance control for nonlinear switched time-delay systems with unmodeled dynamics. *IEEE Transactions on Fuzzy Systems, 26*(4), 1934–1945.

Hua, C., Liu, P. X., & Guan, X. (2009). Backstepping control for nonlinear systems with time delays and applications to chemical reactor systems. *IEEE Transactions on Industrial Electronics, 56*(9), 3723–3732.

Hua, C., Ning, P., & Li, K. (2021). Adaptive prescribed-time control for a class of uncertain nonlinear systems. *IEEE Transactions on Automatic Control, 67*(11), 6159–6166.

Hua, C., Wang, Q.-G., & Guan, X. (2008a). Robust adaptive controller design for nonlinear time-delay systems via T–S fuzzy approach. *IEEE Transactions on Fuzzy Systems, 17*(4), 901–910.

Hua, C., Zhang, J., Luo, X., et al. (2022). Position-velocity constrained trajectory tracking control for unmanned underwater vehicle with model uncertainties. *Ocean Engineering, 266*(2), 112784.

Hua, C., Zhang, L., & Guan, X. (2017b). Distributed adaptive neural network output tracking of leader-following high-order stochastic nonlinear multiagent systems with unknown dead-zone input. *IEEE Transactions on Cybernetics, 47*(1), 177–185.

Hua, C.-C., Wang, Q.-G., & Guan, X.-P. (2008b). Adaptive tracking controller design of nonlinear systems with time delays and unknown dead-zone input. *IEEE Transactions on Automatic Control, 53*(7), 1753–1759.

Huang, Q., Sun, J., & Zhang, C. (2024). High-order fully actuated system approach to robust control of impulsive systems. *IEEE Transactions on Circuits and Systems II: Express Briefs, 71*(3), 1321–1325.

Huang, X., Lin, W., & Yang, B. (2005). Global finite-time stabilization of a class of uncertain nonlinear systems. *Automatica, 41*(5), 881–888.

Ibrir, S. (2011). Observer-based control of a class of time-delay nonlinear systems having triangular structure. *Automatica, 47*(2), 388–394.

Jain, A. K., & Bhasin, S. (2015). Adaptive tracking control of uncertain nonlinear systems with unknown input delay. In *2015 IEEE Conference on Control Applications (CCA)* (pp. 1686–1691)

Jia, F., Lu, J., & Li, Y. (2022). Global output feedback stabilization control for non-strict feedback nonlinear systems. *European Journal of Control, 63*, 126–132.

Jiang, B., Li, C., & Ma, G. (2017). Finite-time output feedback attitude control for spacecraft using "adding a power integrator" technique. *Aerospace Science and Technology, 66*, 342–354.

Jiang, Y., & Jiang, Z.-P. (2017). *Robust adaptive dynamic programming*. Wiley.

Jiang, Z.-P. (2000). Decentralized and adaptive nonlinear tracking of large-scale systems via output feedback. *IEEE Transactions on Automatic Control, 45*(11), 2122–2128.

Jin, T., Li, P., Chen, C., & Du, X. (2010). Fuzzy state feedback stabilization for the time-delay nonlinear system. In *Proceedings of the 29th Chinese Control Conference* (pp. 488–491)

Jin, X. (2017). Adaptive fault tolerant tracking control for a class of stochastic nonlinear systems with output constraint and actuator faults. *Systems and Control Letters, 107*, 100–109.

Jin, X. (2018). Adaptive decentralized finite-time output tracking control for MIMO interconnected nonlinear systems with output constraints and actuator faults. *International Journal of Robust and Nonlinear Control, 28*(5), 1808–1829.

Jin, Z., Qin, Z., Zhang, X., & Guan, C. (2022). A leader-following consensus problem via a distributed observer and fuzzy input-to-output small-gain theorem. *IEEE Transactions on Control of Network Systems, 9*(1), 62–74.

Kanellakopoulos, I., Kokotovic, P. V., & Morse, A. S. (1991). Systematic design of adaptive controllers for feedback linearizable systems. In *1991 American Control Conference* (pp. 649–654). IEEE.

Kang, S., Liu, P. X., & Wang, H. (2022). Fixed-time adaptive fuzzy command filtering control for a class of uncertain nonlinear systems with input saturation and dead zone. *Nonlinear Dynamics, 110*(3), 2401–2414.

Karafyllis, I., & Krstic, M. (2011). Nonlinear stabilization under sampled and delayed measurements, and with inputs subject to delay and zero-order hold. *IEEE Transactions on Automatic Control, 57*(5), 1141–1154.

Khalil, H. K., & Grizzle, J. W. (2002). *Nonlinear systems* (vol. 3). Prentice Hall, Upper Saddle River.

Khattak, A. A., & Iqbal, N. (2006). Design of a robust dynamic state feedback controller for systems with time-varying state delays. In *2006 IEEE International Multitopic Conference* (pp. 451–455)

Kim, B. S., & Yoo, S. J. (2015). Adaptive control of nonlinear pure-feedback systems with output constraints: Integral barrier Lyapunov functional approach. *International Journal of Control, Automation and Systems, 13*, 249–256.

Krishnamurthy, P., Khorrami, F., & Krstic, M. (2020a). A dynamic high-gain design for prescribed-time regulation of nonlinear systems. *Automatica, 115*, 108860.

Krishnamurthy, P., Khorrami, F., & Krstic, M. (2020b). Robust adaptive prescribed-time stabilization via output feedback for uncertain nonlinear strict-feedback-like systems. *European Journal of Control, 55*, 14–23.

Krishnamurthy, P., Khorrami, F., & Krstic, M. (2021). Adaptive output-feedback stabilization in prescribed time for nonlinear systems with unknown parameters coupled with unmeasured states. *International Journal of Adaptive Control and Signal Processing, 35*(2), 184–202.

Krstic, M. (2008). On compensating long actuator delays in nonlinear control. *IEEE Transactions on Automatic Control, 53*(7), 1684–1688.

Krstic, M. (2009). Input delay compensation for forward complete and strict-feedforward nonlinear systems. *IEEE Transactions on Automatic Control, 55*(2), 287–303.

Krstić, M., Kanellakopoulos, I., & Kokotović, P. (1992). Adaptive nonlinear control without overparametrization. *Systems & Control Letters, 19*(3), 177–185.

Krstic, M., Kokotovic, P. V., & Kanellakopoulos, I. (1995). *Nonlinear and adaptive control design*. Wiley.

Li, D., Li, D., Liu, Y., et al. (2017a). Approximation-based adaptive neural tracking control of nonlinear MIMO unknown time-varying delay systems with full state constraints. *IEEE Transactions on Cybernetics, 47*(10), 3100–3109.

Li, H., Bai, L., Zhou, Q., Lu, R., & Wang, L. (2017b). Adaptive fuzzy control of stochastic nonstrict-feedback nonlinear systems with input saturation. *IEEE Transactions on Systems, Man, and Cybernetics: Systems, 47*(8), 2185–2197.

Li, H., Hua, C., Li, K., & Li, Q. (2024a). Adaptive state-quantized control for mismatched nonlinear systems via a dynamic gain approach. *IEEE Transactions on Systems, Man, and Cybernetics: Systems, 54*(3), 1880–1889.

Li, H., Zhou, B., Michiels, W., & Duan, G.-R. (2023a). Prescribed-time unknown input observers design by using periodic delayed output with application to fault estimation. *IEEE Transactions on Systems, Man, and Cybernetics: Systems, 53*(2), 664–674.

Li, J., Du, J., & Chen, C. (2022a). Command-filtered robust adaptive NN control with the prescribed performance for the 3-D trajectory tracking of underactuated Auvs. *IEEE Transactions on Neural Networks and Learning Systems, 33*(11), 6545–6557.

Li, J., Xu, W., Wu, Z., & Liu, Y. (2022b). Practical tracking control via switching for uncertain nonlinear systems with both dead-zone input and output constraint. *IEEE Transactions on Circuits and Systems II: Express Briefs, 70*(6), 2031–2035.

Li, K., Hua, C.-C., You, X., & Guan, X.-P. (2020a). Distributed output-feedback consensus control of multiagent systems with unknown output measurement sensitivity. *IEEE Transactions on Automatic Control, 66*(7), 3303–3310.

Li, M., Li, Y., Ge, S. S., & Lee, T. H. (2016). Adaptive control of robotic manipulators with unified motion constraints. *IEEE Transactions on Systems, Man, and Cybernetics: Systems, 47*(1), 184–194.

Li, M. Y., & Shuai, Z. (2010). Global-stability problem for coupled systems of differential equations on networks. *Journal of Differential Equations, 248*(1), 1–20.

Li, P., & Duan, G. (2023). High-order fully actuated control approaches of flexible servo systems based on singular perturbation theory. *IEEE/ASME Transactions on Mechatronics, 28*(6), 3386–3397.

Li, Q., Hua, C., Li, K., et al. (2024b). A dynamic-event approach to adaptive asymptotic tracking control of p-normal nonlinear systems under full state constraints. *Journal of the Franklin Institute, 361*(1), 357–373.

Li, W., & Krstic, M. (2021). Stochastic nonlinear prescribed-time stabilization and inverse optimality. *IEEE Transactions on Automatic Control, 67*(3), 1179–1193.

Li, W., & Krstic, M. (2022a). Prescribed-time output-feedback control of stochastic nonlinear systems. *IEEE Transactions on Automatic Control, 68*(3), 1431–1446.

Li, W., & Krstic, M. (2022b). Stochastic nonlinear prescribed-time stabilization and inverse optimality. *IEEE Transactions on Automatic Control, 67*(3), 1179–1193.

Li, W., & Krstic, M. (2023). Prescribed-time mean-nonovershooting control under finite-time vanishing noise. *SIAM Journal on Control and Optimization, 61*(3), 1187–1212.

Li, X.-J., & Yang, G.-H. (2017). Adaptive decentralized control for a class of interconnected nonlinear systems via backstepping approach and graph theory. *Automatica, 76*, 87–95.

Li, Y., Liu, Y., & Tong, S. (2022c). Observer-based neuro-adaptive optimized control of strict-feedback nonlinear systems with state constraints. *IEEE Transactions on Neural Networks and Learning Systems, 33*(7), 3131–3145.

Li, Y., Shao, X., & Tong, S. (2020b). Adaptive fuzzy prescribed performance control of nontriangular structure nonlinear systems. *IEEE Transactions on Fuzzy Systems, 28*(10), 2416–2426.

Li, Y., & Tong, S. (2017). Adaptive neural networks decentralized ftc design for nonstrict-feedback nonlinear interconnected large-scale systems against actuator faults. *IEEE Transactions on Neural Networks and Learning Systems, 28*(11), 2541–2554.

Li, Y., & Tong, S. (2018). Adaptive fuzzy control with prescribed performance for block-triangular-structured nonlinear systems. *IEEE Transactions on Fuzzy Systems, 26*(3), 1153–1163.

Li, Y., Tong, S., Liu, L., et al. (2017c). Adaptive output-feedback control design with prescribed performance for switched nonlinear systems. *Automatica, 80*, 225–231.

Li, Y.-X. (2020). Barrier Lyapunov function-based adaptive asymptotic tracking of nonlinear systems with unknown virtual control coefficients. *Automatica, 121*, 109181.

Li, Y.-X., Hou, Z., Che, W.-W., & Wu, Z.-G. (2022d). Event-based design of finite-time adaptive control of uncertain nonlinear systems. *IEEE Transactions on Neural Networks and Learning Systems, 33*(8), 3804–3813.

Li, Y.-X., Wang, Q.-Y., & Tong, S. (2021). Fuzzy adaptive fault-tolerant control of fractional-order nonlinear systems. *IEEE Transactions on Systems, Man, and Cybernetics: Systems, 51*(3), 1372–1379.

Li, Y.-X., & Yang, G.-H. (2016). Adaptive asymptotic tracking control of uncertain nonlinear systems with input quantization and actuator faults. *Automatica, 72*, 177–185.

Li, Z., Cao, G., Xie, W., Gao, R., & Zhang, W. (2023b). Switched-observer-based adaptive neural networks tracking control for switched nonlinear time-delay systems with actuator saturation. *Information Sciences, 621*, 36–57.

Li, Z., Zhang, Y., & Zhang, R. (2022e). Prescribed error performance control for second-order fully actuated systems. *Journal of Systems Science and Complexity, 35*(2), 660–669.

Liang, H., Fu, Y., Gao, J., et al. (2021). Finite-time velocity-observed based adaptive output-feedback trajectory tracking formation control for underactuated unmanned underwater vehicles with prescribed transient performance. *Ocean Engineering, 233*(2), 109071.

Liang, H., Zhang, Y., Huang, T., et al. (2020). Prescribed performance cooperative control for multiagent systems with input quantization. *IEEE Transactions on Cybernetics, 50*(5), 1810–1819.

Lin, W., & Zhang, X. (2019). A dynamic feedback framework for control of time-delay nonlinear systems with unstable zero dynamics. *IEEE Transactions on Automatic Control, 65*(8), 3317–3332.

Lin, Z., & Fang, H. (2007). On asymptotic stabilizability of linear systems with delayed input. *IEEE Transactions on Automatic Control, 52*(6), 998–1013.

Ling, S., Wang, H., & Liu, P. X. (2020). Adaptive tracking control of high-order nonlinear systems under asymmetric output constraint. *Automatica, 122*, 109281.

Liu, C., Liu, X., Wang, H., Zhou, Y., & Gao, C. (2023a). Adaptive control for unknown HOFA nonlinear systems without overparametrization. *International Journal of Robust and Nonlinear Control, 33*(6), 3640–3660.

Liu, C., & Liu, Y. (2023a). Adaptive finite-time stabilization for uncertain nonlinear systems with unknown control coefficients. *Automatica, 149*, 110845.

Liu, C., & Liu, Y. (2023b). Conditions and synthesis of adaptive prescribed-time controllers. *IEEE Transactions on Automatic Control, 68*(12), 8005–8012.

Liu, G. (2022a). Predictive control of high-order fully actuated nonlinear systems with time-varying delays. *Journal of Systems Science and Complexity, 35*(2), 457–470.

Liu, G., Zhang, K., & Li, B. (2022a). Fully-actuated system approach based optimal attitude tracking control of rigid spacecraft with actuator saturation. *Journal of Systems Science and Complexity, 35*(2), 688–702.

Liu, G.-P. (2022b). Coordination of networked nonlinear multi-agents using a high-order fully actuated predictive control strategy. *IEEE/CAA Journal of Automatica Sinica, 9*(4), 615–623.

Liu, H., Wang, Y., & Lewis, F. (2021a). Robust distributed formation controller design for a group of unmanned underwater vehicles. *IEEE Transactions on Systems, Man, and Cybernetics: Systems, 51*(2), 1215–1223.

Liu, J., Zhao, M., & Qiao, L. (2022b). Adaptive barrier Lyapunov function-based obstacle avoidance control for an autonomous underwater vehicle with multiple static and moving obstacles. *Ocean Engineering, 243*, 110303.

Liu, L., Cui, Y., Liu, Y.-J., & Tong, S. (2022c). Adaptive event-triggered output feedback control for nonlinear switched systems based on full state constraints. *IEEE Transactions on Circuits and Systems II: Express Briefs, 69*(9), 3779–3783.

Liu, P.-L., & Su, T.-J. (1998). Robust stability of interval time-delay systems with delay-dependence. *Systems and Control Letters, 33*(4), 231–239.

Liu, S., Su, C.-Y., & Li, Z. (2014). Robust adaptive inverse control of a class of nonlinear systems with prandtl-ishlinskii hysteresis model. *IEEE Transactions on Automatic Control, 59*(8), 2170–2175.

Liu, W., Duan, G., & Hou, M. (2023b). Concurrent learning adaptive command filtered backstepping control for high-order strict-feedback systems. *IEEE Transactions on Circuits and Systems I: Regular Papers, 70*(4), 1696–1709.

Liu, W., Duan, G., & Hou, M. (2023c). High-order command filtered adaptive backstepping control for second- and high-order fully actuated strict-feedback systems. *Journal of the Franklin Institute, 360*(6), 3989–4015.

Liu, X., Chen, M., Sheng, L., & Zhou, D. (2022e). Adaptive fault-tolerant control for nonlinear high-order fully-actuated systems. *Neurocomputing, 495*, 75–85.

Liu, X., Zhang, H., Sun, J., & Guo, X. (2024a). Dynamic threshold finite-time prescribed performance control for nonlinear systems with dead-zone output. *IEEE Transactions on Cybernetics, 54*(1), 655–664.

Liu, X., & Zhang, K. (2019). Input-to-state stability of time-delay systems with delay-dependent impulses. *IEEE Transactions on Automatic Control, 65*(4), 1676–1682.

Liu, X., Zhang, M., Yao, F., et al. (2021b). Barrier Lyapunov function based adaptive region tracking control for underwater vehicles with thruster saturation and dead zone. *Journal of the Franklin Institute, 358*(11), 5820–5844.

Liu, Y., Chen, X., Wu, Y., Cai, H., & Yokoi, H. (2022f). Adaptive neural network control of a flexible spacecraft subject to input nonlinearity and asymmetric output constraint. *IEEE Transactions on Neural Networks and Learning Systems, 33*(11), 6226–6234.

Liu, Y., Liu, J., & Yu, J. (2024b). Singularity avoidance fixed-time adaptive neural control for autonomous underwater vehicles considering unmodelled dynamics and disturbances. *IEEE Transactions on Circuits and Systems II: Express Briefs, 71*(2), 822–826.

Liu, Y., Liu, X., & Jing, Y. (2019). Adaptive fuzzy finite-time stability of uncertain nonlinear systems based on prescribed performance. *Fuzzy Sets and Systems, 374*, 23–39. Theme: Control Engineering.

Liu, Y., Ma, H., & Ma, H. (2018). Adaptive fuzzy fault-tolerant control for uncertain nonlinear switched stochastic systems with time-varying output constraints. *IEEE Transactions on Fuzzy Systems, 26*(5), 2487–2498.

Liu, Y., & Tong, S. (2016). Barrier Lyapunov functions-based adaptive control for a class of nonlinear pure-feedback systems with full state constraints. *Automatica, 64*, 70–75.

Liu, Y., Zhu, Q., Zhao, N., & Wang, L. (2021c). Adaptive fuzzy backstepping control for nonstrict feedback nonlinear systems with time-varying state constraints and backlash-like hysteresis. *Information Sciences, 574*, 606–624.

Liu, Y.-J., & Tong, S. (2017). Barrier Lyapunov functions for nussbaum gain adaptive control of full state constrained nonlinear systems. *Automatica, 76*, 143–152.

Liu, Y.-J., Tong, S., Chen, C. L. P., & Li, D.-J. (2017). Adaptive NN control using integral barrier Lyapunov functionals for uncertain nonlinear block-triangular constraint systems. *IEEE Transactions on Cybernetics, 47*(11), 3747–3757.

Liu, Y.-J., Zhao, W., Liu, L., Li, D., Tong, S., & Chen, C. P. (2021d). Adaptive neural network control for a class of nonlinear systems with function constraints on states. *IEEE Transactions on Neural Networks and Learning Systems, 34*(6), 2732–2741.

Lu, K., Liu, Z., Lai, G., Zhang, Y., & Chen, C. L. P. (2019). Adaptive fuzzy tracking control of uncertain nonlinear systems subject to actuator dead zone with piecewise time-varying parameters. *IEEE Transactions on Fuzzy Systems, 27*(7), 1493–1505.

Lu, S., Tsakalis, K., & Chen, Y. (2023). Development and application of a novel high-order fully actuated system approach–part I: 3-DOF quadrotor control. *IEEE Control Systems Letters, 7,* 1177–1182.

Luo, D., Wang, Y., & Song, Y. (2024). Practical prescribed time tracking control with bounded time-varying gain under non-vanishing uncertainties. *IEEE CAA Journal of Automatica Sinica, 11*(1), 219–230.

Ma, H.-J., & Xu, L.-X. (2021). Decentralized adaptive fault-tolerant control for a class of strong interconnected nonlinear systems via graph theory. *IEEE Transactions on Automatic Control, 66*(7), 3227–3234.

Ma, J., Xu, S., Ma, Q., & Zhang, Z. (2020). Event-triggered adaptive neural network control for nonstrict-feedback nonlinear time-delay systems with unknown control directions. *IEEE Transactions on Neural Networks and Learning Systems, 31*(10), 4196–4205.

Ma, M., Wang, T., Qiu, J., & Karimi, H. R. (2021). Adaptive fuzzy decentralized tracking control for large-scale interconnected nonlinear networked control systems. *IEEE Transactions on Fuzzy Systems, 29*(10), 3186–3191.

Ma, Y., Zhang, K., & Jiang, B. (2023). Prescribed-time fault-tolerant control for fully actuated heterogeneous multiagent systems: A hierarchical design approach. *IEEE Transactions on Aerospace and Electronic Systems, 59*(5), 6624–6636.

Manivannan, R., Samidurai, R., Cao, J., & Perc, M. (2018). Design of resilient reliable dissipativity control for systems with actuator faults and probabilistic time-delay signals via sampled-data approach. *IEEE Transactions on Systems, Man, and Cybernetics: Systems, 50*(11), 4243–4255.

Mao, B., Wu, X., Fan, Z., Lü, J., & Chen, G. (2025). Performance-guaranteed finite-time tracking of strict-feedback systems with unknown control directions: A novel switching mechanism. *IEEE Transactions on Automatic Control, 70,* 4061–4068.

Martin, S., & Whitcomb, L. (2018). Nonlinear model-based tracking control of underwater vehicles with three degree-of-freedom fully coupled dynamical plant models: Theory and experimental evaluation. *IEEE Transactions on Control Systems Technology, 26*(2), 404–414.

Mazenc, F., & Bliman, P.-A. (2006). Backstepping design for time-delay nonlinear systems. *IEEE Transactions on Automatic Control, 51*(1), 149–154.

Mazenc, F., Mondié, S., & Niculescu, S.-I. (2003). Global asymptotic stabilization for chains of integrators with a delay in the input. *IEEE Transactions on Automatic Control, 48*(1), 57–63.

Mazenc, F., Niculescu, S.-I., & Krstic, M. (2012). Lyapunov–Krasovskii functionals and application to input delay compensation for linear time-invariant systems. *Automatica, 48*(7), 1317–1323.

Mechali, O., Xu, L., Xie, X., & Iqbal, J. (2022). Theory and practice for autonomous formation flight of quadrotors via distributed robust sliding mode control protocol with fixed-time stability guarantee. *Control Engineering Practice, 123,* 105150.

Meng, R., Hua, C., Li, K., & Ning, P. (2022a). Adaptive event-triggered control for uncertain high-order fully actuated system. *IEEE Transactions on Circuits and Systems II: Express Briefs, 69*(11), 4438–4442.

Meng, R., Hua, C., Li, K., & Ning, P. (2022b). A multifilters approach to adaptive event-triggered control of uncertain nonlinear systems with global output constraint. *IEEE Transactions on Cybernetics, 54*(2), 1143–1153.

Meng, W., Yang, Q., Si, J., & Sun, Y. (2016). Adaptive neural control of a class of output-constrained nonaffine systems. *IEEE Transactions on Cybernetics, 46*(1), 85–95.

Meng, W., Yang, Q., & Sun, Y. (2015). Adaptive neural control of nonlinear MIMO systems with time-varying output constraints. *IEEE Transactions on Neural Networks and Learning Systems, 26*(5), 1074–1085.

Min, H., Xu, S., Zhang, B., Ma, Q., & Yuan, D. (2022). Fixed-time Lyapunov criteria and state-feedback controller design for stochastic nonlinear systems. *IEEE/CAA Journal of Automatica Sinica, 9*(6), 1005–1014.

Mohan, S., & Kim, J. (2015). Coordinated motion control in task space of an autonomous underwater vehicle-manipulator system. *Ocean Engineering, 104*, 155–167.

Mondal, S., Ghommam, J., & Saad, M. (2018). Homogeneous finite-time consensus control for higher-order multi-agent systems by full order sliding mode. *Journal of Systems Science and Complexity, 31*(5), 1186–1205.

Moradvandi, A., Malek, S., & Shahrokhi, M. (2020). Adaptive finite-time fault-tolerant controller for a class of uncertain mimo nonlinear switched systems subject to output constraints and unknown input nonlinearities. *Nonlinear Analysis: Hybrid Systems, 35*, 100821.

Na, J., Mahyuddin, M. N., Herrmann, G., Ren, X., & Barber, P. (2015). Robust adaptive finite-time parameter estimation and control for robotic systems. *International Journal of Robust and Nonlinear Control, 25*(16), 3045–3071.

Ngo, K., Mahony, R., & Jiang, Z.-P. (2005). Integrator backstepping using barrier functions for systems with multiple state constraints. In *Proceedings of the 44th IEEE Conference on Decision and Control* (pp. 8306–8312)

Nguyen, K.-D., & Dankowicz, H. (2015). Adaptive control of underactuated robots with unmodeled dynamics. *Robotics and Autonomous Systems, 64*, 84–99.

Ni, J., & Shi, P. (2021). Adaptive neural network fixed-time leader–follower consensus for multiagent systems with constraints and disturbances. *IEEE Transactions on Cybernetics, 51*(4), 1835–1848.

Ning, P., Hua, C., Li, K., & Meng, R. (2022a). Adaptive fixed-time control for uncertain nonlinear cascade systems by dynamic feedback. *IEEE Transactions on Systems, Man, and Cybernetics: Systems, 53*(5), 2961–2970.

Ning, P., Hua, C., & Meng, R. (2022b). Adaptive control for a class of nonlinear time-delay system based on the fully actuated system approaches. *Journal of Systems Science and Complexity, 35*(2), 522–534.

Ning, P., Hua, C., Li, K., & Meng, R. (2023a). Event-triggered control for nonlinear uncertain systems via a prescribed-time approach. *IEEE Transactions on Automatic Control, 68*(11), 6975–6981.

Ning, P.-J., Hua, C.-C., Li, K., & Li, H. (2023b). A novel theorem for prescribed-time control of nonlinear uncertain time-delay systems. *Automatica, 152*, 111009.

Ning, P.-J., Hua, C.-C., Li, K., & Meng, R. (2023c). Event-triggered adaptive prescribed-time control for nonlinear systems with uncertain time-varying parameters. *Automatica, 157*, 111229.

Niu, B., Liu, J., Wang, D., Zhao, X., & Wang, H. (2022). Adaptive decentralized asymptotic tracking control for large-scale nonlinear systems with unknown strong interconnections. *IEEE/CAA Journal of Automatica Sinica, 9*(1), 173–186.

Niu, B., & Zhao, J. (2013). Barrier Lyapunov functions for the output tracking control of constrained nonlinear switched systems. *Systems and Control Letters, 62*(10), 963–971.

Orlov, Y. (2004). Finite time stability and robust control synthesis of uncertain switched systems. *SIAM Journal on Control and Optimization, 43*(4), 1253–1271.

Orlov, Y. (2020). *Nonsmooth Lyapunov analysis in finite and infinite dimensions*. Springer.

Orlov, Y. (2022). Time space deformation approach to prescribed-time stabilization: Synergy of time-varying and non-lipschitz feedback designs. *Automatica, 144*, 110485.

Orlov, Y., & Krstic, M. (2022). Comments on "design of controllers with arbitrary convergence time" [automatica 112 (2020) 108710]. *Automatica, 142*, 110429.

Ouyang, H., & Lin, Y. (2021). Adaptive fault-tolerant control and performance recovery against actuator failures with deferred actuator replacement. *IEEE Transactions on Automatic Control, 66*(8), 3810–3817.

Pal, A. K., Kamal, S., Nagar, S. K., Bandyopadhyay, B., & Fridman, L. (2020). Design of controllers with arbitrary convergence time. *Automatica, 112*, 108710.

Pan, Y., Ji, W., Lam, H., et al. (2024). An improved predefined-time adaptive neural control approach for nonlinear multiagent systems. *IEEE Transactions on Automation Science and Engineering, 21*(4), 6311–6320.

Peng, K., Wang, H., Zhang, H., et al. (2023). Multivariable decoupling control of civil turbofan engines based on fully actuated system approach. *Journal of Systems Science and Complexity, 36*(3), 947–959.

Perez-Arancibia, N. O., Tsao, T.-C., & Gibson, J. S. (2010). Saturation-induced instability and its avoidance in adaptive control of hard disk drives. *IEEE Transactions on Control Systems Technology, 18*(2), 368–382.

Peterka, V. (1999). Adaptive digital regulation of noisy systems. *International Journal of Adaptive Control and Signal Processing, 13*(6), 537–550.

Pishro, A., Shahrokhi, M., & Sadeghi, H. (2022). Fault-tolerant adaptive fractional controller design for incommensurate fractional-order nonlinear dynamic systems subject to input and output restrictions. *Chaos, Solitons and Fractals, 157*, 111930.

Plestan, F., Shtessel, Y., Bregeault, V., & Poznyak, A. (2010). New methodologies for adaptive sliding mode control. *International Journal of Control, 83*(9), 1907–1919.

Polyakov, A. (2012). Nonlinear feedback design for fixed-time stabilization of linear control systems. *IEEE Transactions on Automatic Control, 57*(8), 2106–2110.

Polyakov, A. (2020). *Generalized homogeneity in systems and control*. Springer Nature Link (pp. 2197–7119). Springer.

Pongvuthithum, R., Rattanamongkhonkun, K., & Lin, W. (2017). Asymptotic regulation of time-delay nonlinear systems with unknown control directions. *IEEE Transactions on Automatic Control, 63*(5), 1495–1502.

Qiu, Y., Liang, X., Dai, Z., Cao, J., & Chen, Y. (2015). Backstepping dynamic surface control for a class of non-linear systems with time-varying output constraints. *IET Control Theory & Applications, 9*(15), 2312–2319.

Ren, B., Ge, S. S., Tee, K. P., & Lee, T. H. (2010). Adaptive neural control for output feedback nonlinear systems using a barrier lyapunov function. *IEEE Transactions on Neural Networks, 21*(8), 1339–1345.

Ricardo, J. A., Jr., & Santos, D. A. (2022). Smooth second-order sliding mode control for fully actuated multirotor aerial vehicles. *ISA Transactions, 129*, 169–178.

Roy, T. K., Mahmud, M. A., Shen, W., & Oo, A. M. T. (2016). Nonlinear adaptive excitation controller design for multimachine power systems with unknown stability sensitive parameters. *IEEE Transactions on Control Systems Technology, 25*(6), 2060–2072.

Ruan, Z., Yang, Q., Ge, S., et al. (2022). Adaptive fuzzy fault tolerant control of uncertain mimo nonlinear systems with output constraints and unknown control directions. *IEEE Transactions on Fuzzy Systems, 30*(5), 1224–1238.

Sastry, S., & Bodson, M. (2011). *Adaptive control: Stability, convergence and robustness*. Courier Corporation.

Sastry, S. S., & Isidori, A. (1989). Adaptive control of linearizable systems. *IEEE Transactions on Automatic Control, 34*(11), 1123–1131.

Shao, X., Tian, B., & Yang, W. (2021). Fixed-time trajectory following for quadrotors via output feedback. *ISA Transactions, 110*, 213–224.

Shao, X., & Ye, D. (2022). Neuroadaptive deferred full-state constraints control without feasibility conditions for uncertain nonlinear easss. *Journal of the Franklin Institute, 359*(7), 2810–2832.

Shen, Q., Shi, P., Shi, Y., et al. (2017). Adaptive output consensus with saturation and dead-zone and its application. *IEEE Transactions on Industrial Electronics, 64*(6), 5025–5034.

Shen, Y., & Huang, Y. (2012). Global finite-time stabilisation for a class of nonlinear systems. *International Journal of Systems Science, 43*(1), 73–78.

Shen, Y., & Xia, X. (2008). Semi-global finite-time observers for nonlinear systems. *Automatica, 44*(12), 3152–3156.

Shi, X.-N., Zhou, Z.-G., Zhou, D., & Li, R. (2020). Event-triggered fixed-time adaptive trajectory tracking for a class of uncertain nonlinear systems with input saturation. *IEEE Transactions on Circuits and Systems II: Express Briefs, 68*(3), 983–987.

Shi, Y., Xie, W., Xiong, M., et al. (2023). Neural adaptive quantitative prescribed performance sectionalized event-triggered control for autonomous underwater vehicles. *IEEE Transactions on Intelligent Transportation Systems, 24*(10), 10857–10868.

Shojaei, K., & Arefi, M. (2015). On the neuro-adaptive feedback linearising control of underactuated autonomous underwater vehicles in three-dimensional space. *IET Control Theory and Applications, 8*(9), 1264–1273.

Singh, S., & Jain, A. (2024). Collision avoidance and connectivity preservation using asymmetric barrier Lyapunov function with time-varying distance-constraints. *Systems and Control Letters, 183*, 105672.

Sipahi, R., Vyhlídal, T., Niculescu, S.-I., Pepe, P., et al. (2012). *Time delay systems: Methods, applications and new trends.* Springer.

Song, S., Park, J. H., Zhang, B., & Song, X. (2021a). Event-triggered adaptive practical fixed-time trajectory tracking control for unmanned surface vehicle. *IEEE Transactions on Circuits and Systems II: Express Briefs, 68*(1), 436–440.

Song, S., Zhang, B., Song, X., & Zhang, Z. (2021b). Neuro-fuzzy-based adaptive dynamic surface control for fractional-order nonlinear strict-feedback systems with input constraint. *IEEE Transactions on Systems, Man, and Cybernetics: Systems, 51*(6), 3575–3586.

Song, X., Man, J., & Ahn, C. K. (2021c). Joint state and fault estimation for networked interconnected PDE systems with semi-markov fault coefficient via conjunct measurement. *IEEE Transactions on Circuits and Systems I: Regular Papers, 68*(9), 3869–3880.

Song, Y., Wang, Y., Holloway, J., & Krstic, M. (2017). Time-varying feedback for regulation of normal-form nonlinear systems in prescribed finite time. *Automatica, 83*, 243–251.

Song, Y., Wang, Y., & Krstic, M. (2019). Time-varying feedback for stabilization in prescribed finite time. *International Journal of Robust and Nonlinear Control, 29*(3), 618–633.

Song, Y., Ye, H., & Lewis, F. L. (2023). Prescribed-time control and its latest developments. *IEEE Transactions on Systems, Man, and Cybernetics: Systems, 53*(7), 4102–4116.

Song, Y., & Zhou, S. (2018). Tracking control of uncertain nonlinear systems with deferred asymmetric time-varying full state constraints. *Automatica, 98*, 314–322.

Song, Z., Li, P., Wang, Z., Huang, X., & Liu, W. (2020). Adaptive tracking control for switched uncertain nonlinear systems with input saturation and unmodeled dynamics. *IEEE Transactions on Circuits and Systems II: Express Briefs, 67*(12), 3152–3156.

Steeves, D., Krstic, M., & Vazquez, R. (2020). Prescribed–time estimation and output regulation of the linearized schrödinger equation by backstepping. *European Journal of Control, 55*, 3–13. Finite-time estimation, diagnosis and synchronization of uncertain systems.

Su, C.-Y., Wang, Q., Chen, X., & Rakheja, S. (2005). Adaptive variable structure control of a class of nonlinear systems with unknown prandtl-ishlinskii hysteresis. *IEEE Transactions on Automatic Control, 50*(12), 2069–2074.

Su, H., Zhang, H., Liang, X., & Liu, C. (2020). Decentralized event-triggered online adaptive control of unknown large-scale systems over wireless communication networks. *IEEE Transactions on Neural Networks and Learning Systems, 31*(11), 4907–4919.

Su, X., Liu, Z., Lai, G., Chen, C. L. P., & Chen, C. (2017). Direct adaptive compensation for actuator failures and dead-zone constraints in tracking control of uncertain nonlinear systems. *Information Sciences, 417*, 328–343.

Sui, S., Chen, C. L. P., & Tong, S. (2021a). Neural-network-based adaptive DSC design for switched fractional-order nonlinear systems. *IEEE Transactions on Neural Networks and Learning Systems, 32*(10), 4703–4712.

Sui, S., Chen, C. L. P., & Tong, S. (2021b). A novel adaptive nn prescribed performance control for stochastic nonlinear systems. *IEEE Transactions on Neural Networks and Learning Systems, 32*(7), 3196–3205.

Sui, S., Xu, H., Chen, C. L. P., & Tong, S. (2023). Nonsingular fixed-time control of nonstrict feedback MIMO nonlinear system with asymptotically convergent tracking error. *IEEE Transactions on Fuzzy Systems, 31*(5), 1689–1702.

Sun, H., Hou, L., Zong, G., et al. (2020a). Adaptive decentralized neural network tracking control for uncertain interconnected nonlinear systems with input quantization and time delay. *IEEE Transactions on Neural Networks and Learning Systems, 31*(4), 1401–1409.

Sun, W., Su, S.-F., Wu, Y., & Xia, J. (2022). Adaptive fuzzy event-triggered control for high-order nonlinear systems with prescribed performance. *IEEE Transactions on Cybernetics, 52*(5), 2885–2895.

Sun, W., Su, S.-F., Wu, Y., Xia, J., & Nguyen, V.-T. (2020b). Adaptive fuzzy control with high-order barrier Lyapunov functions for high-order uncertain nonlinear systems with full-state constraints. *IEEE Transactions on Cybernetics, 50*(8), 3424–3432.

Sun, W., Wu, Y.-Q., & Sun, Z.-Y. (2020c). Command filter-based finite-time adaptive fuzzy control for uncertain nonlinear systems with prescribed performance. *IEEE Transactions on Fuzzy Systems, 28*(12), 3161–3170.

Sun, Y., Li, C., Qin, H., et al. (2021). Robust neural network-based tracking control for unmanned surface vessels under deferred asymmetric constraints. *International Journal of Robust and Nonlinear Control, 32*(5), 2741–2759.

Tabuada, P. (2007). Event-triggered real-time scheduling of stabilizing control tasks. *IEEE Transactions on Automatic Control, 52*(9), 1680–1685.

Tang, Z.-L., Ge, S. S., Tee, K. P., & He, W. (2016). Robust adaptive neural tracking control for a class of perturbed uncertain nonlinear systems with state constraints. *IEEE Transactions on Systems, Man, and Cybernetics: Systems, 46*(12), 1618–1629.

Tao, G., & Kokotovic, P. (1995). Adaptive control of plants with unknown hystereses. *IEEE Transactions on Automatic Control, 40*(2), 200–212.

Tee, K. P., & Ge, S. S. (2012). Control of state-constrained nonlinear systems using integral barrier Lyapunov functionals. In *2012 IEEE 51st IEEE Conference on Decision and Control (CDC)* (pp. 3239–3244)

Tee, K. P., Ge, S. S., & Tay, E. H. (2009). Barrier Lyapunov functions for the control of output-constrained nonlinear systems. *Automatica, 45*(4), 918–927.

Tee, K. P., Ren, B., & Ge, S. S. (2011). Control of nonlinear systems with time-varying output constraints. *Automatica, 47*(11), 2511–2516.

Teel, A., Kadiyala, R., Kokotovic, P., & Sastry, S. (1991). Indirect techniques for adaptive input-output linearization of non-linear systems. *International Journal of Control, 53*(1), 193–222.

Tian, B., Lu, H., Zuo, Z., & Wang, H. (2018). Fixed-time stabilization of high-order integrator systems with mismatched disturbances. *Nonlinear Dynamics, 94*, 2889–2899.

Tian, B., Zuo, Z., Yan, X., & Wang, H. (2017). A fixed-time output feedback control scheme for double integrator systems. *Automatica, 80*, 17–24.

Tian, D., & Song, X. (2023). Addressing complex state constraints in the integral barrier Lyapunov function-based adaptive tracking control. *International Journal of Control, 96*(5), 1202–1209.

Tian, G., Li, B., Zhao, Q., & Duan, G. (2024). High-precision trajectory tracking control for free-flying space manipulators with multiple constraints and system uncertainties. *IEEE Transactions on Aerospace and Electronic Systems, 60*(1), 789–801.

Tong, S., Huo, B., & Li, Y. (2014). Observer-based adaptive decentralized fuzzy fault-tolerant control of nonlinear large-scale systems with actuator failures. *IEEE Transactions on Fuzzy Systems, 22*(1), 1–15.

Tong, S., Min, X., & Li, Y. (2020). Observer-based adaptive fuzzy tracking control for strict-feedback nonlinear systems with unknown control gain functions. *IEEE Transactions on Cybernetics, 50*(9), 3903–3913.

Tong, S. C., Li, Y. M., & Zhang, H.-G. (2011). Adaptive neural network decentralized backstepping output-feedback control for nonlinear large-scale systems with time delays. *IEEE Transactions on Neural Networks, 22*(7), 1073–1086.

Tran, D., & Yucelen, T. (2020). Finite-time control of perturbed dynamical systems based on a generalized time transformation approach. *Systems & Control Letters, 136*, 104605.

Wang, A., Liu, L., Qiu, J., & Feng, G. (2022a). Event-triggered adaptive fuzzy output-feedback control for nonstrict-feedback nonlinear systems with asymmetric output constraint. *IEEE Transactions on Cybernetics, 52*(1), 712–722.

Wang, C., & Lin, Y. (2015). Decentralized adaptive tracking control for a class of interconnected nonlinear time-varying systems. *Automatica, 54*, 16–24.

Wang, C., Liu, X., & Wang, H. (2021). An adaptive fault-tolerant control scheme for a class of fractional-order systems with unknown input dead-zones. *International Journal of Systems Science, 52*(2), 291–306.

Wang, C.-C., & Yang, G.-H. (2018). Observer-based adaptive prescribed performance tracking control for nonlinear systems with unknown control direction and input saturation. *Neurocomputing, 28*(4), 17–26.

Wang, H., Chen, B., Lin, C., Sun, Y., & Wang, F. (2017). Adaptive finite-time control for a class of uncertain high-order non-linear systems based on fuzzy approximation. *IET Control Theory & Applications, 11*(5), 677–684.

Wang, H., Karimi, H. R., Liu, P. X., & Yang, H. (2018). Adaptive neural control of nonlinear systems with unknown control directions and input dead-zone. *IEEE Transactions on Systems, Man, and Cybernetics: Systems, 48*(11), 1897–1907.

Wang, L., Sun, W., & Su, S.-F. (2022b). Adaptive asymptotic tracking control for nonlinear systems with state constraints and input saturation. *Applied Mathematics and Computation, 431*, 127342.

Wang, L., Wang, M., & Meng, W. (2023a). System transformation-based event-triggered fuzzy control for state constrained nonlinear systems with unknown control directions. *IEEE Transactions on Fuzzy Systems, 31*(7), 2331–2344.

Wang, M., Chen, B., & Shi, P. (2008). Adaptive neural control for a class of perturbed strict-feedback nonlinear time-delay systems. *IEEE Transactions on Systems, Man, and Cybernetics, Part B (Cybernetics), 38*(3), 721–730.

Wang, N., Liu, X., Liu, C., Wang, H., & Zhou, Y. (2023b). Adaptive control and almost disturbance decoupling for uncertain hofa nonlinear systems. *International Journal of Adaptive Control and Signal Processing, 37*(8), 2133–2161.

Wang, N., Liu, X. P., Liu, C. G., Wang, H. Q., & Zhou, Y.-C. (2022c). Almost disturbance decoupling for hofa nonlinear systems with strict-feedback form. *Journal of Systems Science and Complexity, 35*(21), 481–501.

Wang, N., Tao, F., Fu, Z., et al. (2022d). Adaptive fuzzy control for a class of stochastic strict feedback high-order nonlinear systems with full-state constraints. *IEEE Transactions on Systems, Man, and Cybernetics: Systems, 52*(1), 205–213.

Wang, N., & Wang, Y. (2022). Fuzzy adaptive quantized tracking control of switched high-order nonlinear systems: A new fixed-time prescribed performance method. *IEEE Transactions on Circuits and Systems II: Express Briefs, 69*(7), 3279–3283.

Wang, W., Wen, T., He, X., et al. (2023c). Path following with prescribed performance for under-actuated autonomous underwater vehicles subjects to unknown actuator dead-zone. *IEEE Transactions on Intelligent Transportation Systems, 24*(6), 6257–6267.

Wang, X. (2021). Active fault tolerant control for unmanned underwater vehicle with actuator fault and guaranteed transient performance. *IEEE Transactions on Intelligent Vehicles, 6*(3), 470–479.

Wang, X., & Duan, G. (2024). Fully actuated system approaches: Predictive elimination control for discrete-time nonlinear time-varying systems with full state constraints and time-varying delays. *IEEE Transactions on Circuits and Systems I: Regular Papers, 71*(1), 383–396.

Wang, X., Wang, Q., & Sun, C. (2022e). Prescribed performance fault-tolerant control for uncertain nonlinear MIMO system using actor–critic learning structure. *IEEE Transactions on Neural Networks and Learning Systems, 33*(9), 4479–4490.

Wang, Y., & Song, Y. (2019). A general approach to precise tracking of nonlinear systems subject to non-vanishing uncertainties. *Automatica, 106*, 306–314.

Wang, Z., & Yuan, J. (2020). Fuzzy adaptive fault tolerant IGC method for STT missiles with time-varying actuator faults and multisource uncertainties. *Journal of the Franklin Institute, 357*(1):59–81.

Wei, J., Liu, Y.-J., Chen, H., & Liu, L. (2023). Fuzzy adaptive control for vehicular platoons with constraints and unknown dead-zone input. *IEEE Transactions on Intelligent Transportation Systems, 24*(4), 4403–4412.

Weisler, W., Stewart, W., Anderson, M., et al. (2018). Testing and characterization of a fixed wing cross-domain unmanned vehicle operating in aerial and underwater environments. *IEEE Journal of Ocean Engineering, 43*(4), 969–982.

Wen, C. (1994). Decentralized adaptive regulation. *IEEE Transactions on Automatic Control, 39*(10), 2163–2166.

Wen, C., & Zhou, J. (2007). Decentralized adaptive stabilization in the presence of unknown backlash-like hysteresis. *Automatica, 43*(3), 426–440.

Wen, C., Zhou, J., Liu, Z., & Su, H. (2011). Robust adaptive control of uncertain nonlinear systems in the presence of input saturation and external disturbance. *IEEE Transactions on Automatic Control, 56*(7), 1672–1678.

Wen, X., Zou, W., & Xiang, Z. (2023). Finite-time consensus for high-order disturbed multiagent systems with bounded control input. *IEEE Systems Journal, 17*(4), 6380–6389.

Wu, J., Sun, W., Su, S.-F., & Xia, J. (2022). Neural-based adaptive control for nonlinear systems with quantized input and the output constraint. *Applied Mathematics and Computation, 413*, 126637.

Wu, J., Yang, Y., Chen, W., Wang, H., & Wu, Z.-G. (2025). Adaptive neural fault tolerant control for input-delayed stochastic systems subject to states and input quantization. *IEEE Transactions on Systems, Man, and Cybernetics: Systems, 55*(4), 2451–2462.

Wu, X., Sun, W., Su, S.-F., & Xie, X. (2024). Adaptive stabilization for high-order fully actuated systems with unknown control directions. *IEEE Transactions on Systems, Man, and Cybernetics: Systems, 54*(8), 5150–5159.

Wu, Y., & Xie, X.-J. (2021). Robust adaptive control for state-constrained nonlinear systems with input saturation and unknown control direction. *IEEE Transactions on Systems, Man, and Cybernetics: Systems, 51*(2), 1192–1202.

Xia, X., Zhang, T., Kang, G., & Fang, Y. (2022). Adaptive control of uncertain nonlinear systems with discontinuous input and time-varying input delay. *IEEE Transactions on Systems, Man, and Cybernetics: Systems, 52*(11), 7248–7258.

Xiao, F., & Chen, L. (2022). Attitude control of spherical liquid-filled spacecraft based on high-order fully actuated system approaches. *Journal of Systems Science and Complexity, 35*(2), 471–480.

Xiao, F.-Z., & Chen, L.-Q. (2023). Fully actuated systems in terms of quaternions for spacecraft attitude control. *Acta Astronautica, 209*, 1–5.

Xiao, X. B. (2018). Prescribed performance-based low-computational cost fuzzy control of a hypersonic vehicle using non-affine models. *Advances in Mechanical Engineering, 10*(2), 2018.

Xiao, Y., Cai, G., & Duan, G. (2024). High-order adaptive dynamic surface control for output-constrained nonlinear systems based on fully actuated system approach. *International Journal of Systems Science, 55*(3), 482–498.

Xin, C., Li, Y.-X., & Ahn, C. K. (2023). Adaptive neural asymptotic tracking of uncertain non-strict feedback systems with full-state constraints via command filtered technique. *IEEE Transactions on Neural Networks and Learning Systems, 34*(10), 8102–8107.

Xing, L., Wen, C., Liu, Z., et al. (2017). Event-triggered adaptive control for a class of uncertain nonlinear systems. *IEEE Transactions on Automatic Control, 62*(4), 2071–2076.

Xing, L., Wen, C., Zhu, Y., Su, H., & Liu, Z. (2016). Output feedback control for uncertain nonlinear systems with input quantization. *Automatica, 65*, 191–202.

Xu, H., Yu, D., Sui, S., Zhao, Y.-P., Chen, C. L. P., & Wang, Z. (2023). Nonsingular practical fixed-time adaptive output feedback control of MIMO nonlinear systems. *IEEE Transactions on Neural Networks and Learning Systems, 34*(10), 7222–7234.

Xu, J.-X., & Jin, X. (2013). State-constrained iterative learning control for a class of MIMO systems. *IEEE Transactions on Automatic Control, 58*(5), 1322–1327.

Xu, Z., Xie, N., Shen, H., Hu, X., & Liu, Q. (2021). Extended state observer-based adaptive prescribed performance control for a class of nonlinear systems with full-state constraints and uncertainties. *IEEE Transactions on Cybernetics, 105*(1), 345–358.

Xu, X. (2024). High-order fully actuated system models for discrete-time strict-feedback systems with increasing dimensions. *International Journal of Systems Science, 55*(12), 2454–2463.

Xue, H., & Ou, Y. (2023). A novel asymmetric barrier Lyapunov function-based fixed-time ship berthing control under multiple state constraints. *Ocean Engineering, 281*, 114756.

Yang, R., & Wang, Y. (2014). New delay-dependent stability criteria and robust control of nonlinear time-delay systems. *Journal of Systems Science and Complexity, 27*(5), 883–898.

Yang, S., Pan, Y., Cao, L., et al. (2024). Predefined-time fault-tolerant consensus tracking control for multi-UAV systems with prescribed performance and attitude constraints. *IEEE Transactions on Aerospace and Electronic Systems, 60*(4), 4058–4072.

Yang, W., Yu, W., & Zheng, W. X. (2022a). Fault-tolerant adaptive fuzzy tracking control for nonaffine fractional-order full-state-constrained MISO systems with actuator failures. *IEEE Transactions on Cybernetics, 52*(8), 8439–8452.

Yang, X., & Cao, J. (2010). Finite-time stochastic synchronization of complex networks. *Applied Mathematical Modelling, 34*(11), 3631–3641.

Yang, X., Feng, G., He, C., & Cao, J. (2022b). Event-triggered dynamic output quantization control of switched T–S fuzzy systems with unstable modes. *IEEE Transactions on Fuzzy Systems, 30*(10), 4201–4210.

Yang, Z., Zhang, H., & Cui, Y. (2018). Adaptive fuzzy tracking control for switched uncertain strict-feedback nonlinear systems. *Journal of the Franklin Institute, 355*(2), 714–727.

Ye, H., & Song, Y. (2023). Prescribed-time control for linear systems in canonical form via nonlinear feedback. *IEEE Transactions on Systems, Man, and Cybernetics: Systems, 53*(2), 1126–1135.

Yin, J., Khoo, S., & Man, Z. (2017). Finite-time stability theorems of homogeneous stochastic nonlinear systems. *Systems and Control Letters, 100*, 6–13.

Yu, J., Shi, P., & Zhao, L. (2018). Finite-time command filtered backstepping control for a class of nonlinear systems. *Automatica, 92*, 173–180.

Yu, X., & Lin, Y. (2016). Adaptive backstepping quantized control for a class of nonlinear systems. *IEEE Transactions on Automatic Control, 62*(2), 981–985.

Yu, X., Wang, T., & Gao, H. (2020). Adaptive neural fault-tolerant control for a class of strict-feedback nonlinear systems with actuator and sensor faults. *Neurocomputing, 380*, 87–94.

Yu, Y., Liu, G.-P., Huang, Y., & Guerrero, J. M. (2024). Coordinated predictive secondary control for DC microgrids based on high-order fully actuated system approaches. *IEEE Transactions on Smart Grid, 15*(1), 19–33.

Yu, Y., Pei, H.-L., & Yu, J. (2023). Exponential stabilization of a rotating body-beam system with boundary output constraint under unknown time-varying disturbance. *IEEE Transactions on Circuits and Systems II: Express Briefs, 71*(4), 2239–2243.

Yuan, F., Liu, Y.-J., Liu, L., Lan, J., Li, D., Tong, S., & Chen, C. L. P. (2023). Adaptive neural consensus tracking control for nonlinear multiagent systems using integral barrier lyapunov functionals. *IEEE Transactions on Neural Networks and Learning Systems, 34*(8), 4544–4554.

Yuan, M., & Zhai, J. (2022). Dynamic event-triggered control for a class of p-normal nonlinear time-delay systems via output feedback. *International Journal of Robust and Nonlinear Control, 33*, 507–525.

Yuan, X., Chen, B., & Lin, C. (2022). Prescribed finite-time adaptive neural tracking control for nonlinear state-constrained systems: Barrier function approach. *IEEE Transactions on Neural Networks and Learning Systems, 33*(12), 7513–7522.

Yucelen, T., Kan, Z., & Pasiliao, E. (2019). Finite-time cooperative engagement. *IEEE Transactions on Automatic Control, 64*(8), 3521–3526.

Zhan, Y., Li, X., & Tong, S. (2023). Observer-based decentralized control for non-strict-feedback fractional-order nonlinear large-scale systems with unknown dead zones. *IEEE Transactions on Neural Networks and Learning Systems, 34*(10), 7479–7490.

Zhang, C.-H., Wang, L., Zhang, Y., Ding, W., & Hua, C. (2025). High-order fully actuated system approaches for a class of pseudo pure-feedback nonlinear systems. *IEEE Transactions on Systems, Man, and Cybernetics: Systems, 55*(1), 635–644.

Zhang, D., Liu, G., & Cao, L. (2023a). Constrained cooperative control for high-order fully actuated multiagent systems with application to air-bearing spacecraft simulators. *IEEE/ASME Transactions on Mechatronics, 28*(3), 1570–1581.

Zhang, D., Liu, G., & Cao, L. (2023b). Predictive control of discrete-time high-order fully actuated systems with application to air-bearing spacecraft simulator. *Journal of the Franklin Institute, 360*(8), 5910–5927.

Zhang, D., Liu, G., & Cao, L. (2023c). Proportional integral predictive control of high-order fully actuated networked multiagent systems with communication delays. *IEEE Transactions on Systems, Man, and Cybernetics: Systems, 52*(3), 801–812.

Zhang, D.-W., & Liu, G.-P. (2023). Predictive control for networked high-order fully actuated systems subject to communication delays and external disturbances. *ISA Transactions, 139*, 425–435.

Zhang, D.-W., Liu, G.-P., & Cao, L. (2022a). Coordinated control of high-order fully actuated multiagent systems and its application: A predictive control strategy. *IEEE/ASME Transactions on Mechatronics, 27*(6), 4362–4372.

Zhang, D.-W., Liu, G.-P., & Cao, L. (2023d). Proportional integral predictive control of high-order fully actuated networked multiagent systems with communication delays. *IEEE Transactions on Systems, Man, and Cybernetics: Systems, 53*(2), 801–812.

Zhang, J., & Raissi, T. (2021). Indefinite Lyapunov crazumikhin functions-based stability and event-triggered control of switched nonlinear time-delay systems. *IEEE Transactions on Circuits and Systems II: Express Briefs, 68*(10), 3286–3290.

Zhang, J., & Tong, S. (2021). Adaptive fuzzy output feedback FTC for nonstrict-feedback systems with sensor faults and dead zone input. *Neurocomputing, 435*, 67–76.

Zhang, L., Wang, P., & Hua, C. (2023e). Adaptive control of time-delay nonlinear HOFA systems with unmodeled dynamics and unknown dead-zone input. *International Journal of Robust and Nonlinear Control, 33*(4), 2615–2628.

Zhang, L., Zhu, L., & Hua, C. (2022b). Practical prescribed time control based on high-order fully actuated system approach for strong interconnected nonlinear systems. *Nonlinear Dynamics, 110*(4), 3535–3545.

Zhang, L., Zhu, L., Hua, C., & Qian, C. (2021a). Decentralised state-feedback prescribed performance control for a class of interconnected nonlinear full-state time-delay systems with strong interconnection. *International Journal of Systems Science, 52*(12), 2580–2596.

Zhang, L., Zhu, L., Hua, C., et al. (2023f). Adaptive decentralized control for interconnected time-delay uncertain nonlinear systems with different unknown control directions and deferred full-state constraints. *IEEE Transactions on Neural Networks and Learning Systems, 34*(12), 10789–10801.

Zhang, L., Liu, S., & Hua, C. (2022). Truncated predictor stabilization control for interconnected nonlinear systems with time-varying input delay. *Nonlinear Dynamics, 107*(3), 2421–2428.

Zhang, T., & Ge, S. S. (2009). Adaptive neural network tracking control of MIMO nonlinear systems with unknown dead zones and control directions. *IEEE Transactions on Neural Networks, 20*(3), 483–497.

Zhang, T., Hua, Y., Xia, X., & Yi, Y. (2021b). Unified adaptive event-triggered control of uncertain multi-input multi-output nonlinear systems with dynamic and static constraints. *International Journal of Robust and Nonlinear Control, 31*(6), 2371–2392.

Zhang, T., Xia, M., & Yi, Y. (2017a). Adaptive neural dynamic surface control of strict-feedback nonlinear systems with full state constraints and unmodeled dynamics. *Automatica, 81*, 232–239.

Zhang, X., Lin, W., & Lin, Y. (2016). Nonsmooth feedback control of time-delay nonlinear systems: A dynamic gain based approach. *IEEE Transactions on Automatic Control, 62*(1), 438–444.

Zhang, X., Lin, W., & Lin, Y. (2017b). Adaptive regulation of time-delay cascade systems with parameter uncertainty. *IFAC-PapersOnLine, 50*(1), 723–728. 20th IFAC World Congress.

Zhang, X., Lin, W., & Lin, Y. (2018a). Adaptive control of time-delay cascade systems with unknown parameters by partial state feedback. *Automatica, 94*, 45–54.

Zhang, X., Lin, W., & Lin, Y. (2018b). Iterative changing supply rates, dynamic state feedback, and adaptive stabilization of time-delay systems. *IEEE Transactions on Automatic Control, 64*(2), 751–758.

Zhang, X., & Lin, Y. (2014). Nonlinear decentralized control of large-scale systems with strong interconnections. *Automatica, 50*(9), 2419–2423.

Zhang, X., Liu, L., Feng, G., & Zhang, C. (2013). Output feedback control of large-scale nonlinear time-delay systems in lower triangular form. *Automatica, 49*(11), 3476–3483.

Zhang, Z., Duan, G., & Hu, Y. (2018c). Robust adaptive control for a class of semi-strict feedback systems with state and input constraints. *International Journal of Robust and Nonlinear Control, 28*(9), 3189–3211.

Zhang, Z., Gao, Y., Sun, W., & Wu, Y. (2023g). Fixed-time event-triggered controller of a state-constrained nonlinear system: A zero-error tracking control approach. *International Journal of Robust and Nonlinear Control, 33*(15), 9271–9298.

Zhang, Z., Wen, C., Xing, L., & Song, Y. (2022c). Adaptive output feedback control of nonlinear systems with mismatched uncertainties under input/output quantization. *IEEE Transactions on Automatic Control, 67*(9), 4801–4808.

Zhao, C., & Lin, W. (2021). Global stabilization by memoryless feedback for nonlinear systems with a limited input delay and large state delays. *IEEE Transactions on Automatic Control, 66*(8), 3702–3709.

Zhao, J., Yuan, Y., yao Sun, Z., & Xie, X. (2023a). Applications to the dynamics of the suspension system of fast finite time stability in probability of p-norm stochastic nonlinear systems. *Applied Mathematics and Computation, 457*, 128221.

Zhao, K., Chen, L., Meng, W., & Zhao, L. (2023b). Unified mapping function-based neuroadaptive control of constrained uncertain robotic systems. *IEEE Transactions on Cybernetics, 53*(6), 3665–3674.

Zhao, K., & Song, Y. (2019). Removing the feasibility conditions imposed on tracking control designs for state-constrained strict-feedback systems. *IEEE Transactions on Automatic Control, 64*(3), 1265–1272.

Zhao, K., & Song, Y. (2020). Neuroadaptive robotic control under time-varying asymmetric motion constraints: A feasibility-condition-free approach. *IEEE Transactions on Cybernetics, 50*(1), 15–24.

Zhao, K., Song, Y., Chen, C., et al. (2022a). Adaptive asymptotic tracking with global performance for nonlinear systems with unknown control directions. *IEEE Transactions on Automatic Control, 67*(3), 1566–1573.

Zhao, K., Song, Y., & Chen, L. (2019a). Tracking control of nonlinear systems with improved performance via transformational approach. *International Journal of Robust and Nonlinear Control, 29*(6), 1789–1806.

Zhao, K., Song, Y., & Shen, Z. (2018). Neuroadaptive fault-tolerant control of nonlinear systems under output constraints and actuation faults. *IEEE Transactions on Neural Networks and Learning Systems, 29*(2), 286–298.

Zhao, K., Song, Y., & Zhang, Z. (2019b). Tracking control of MIMO nonlinear systems under full state constraints: A single-parameter adaptation approach free from feasibility conditions. *Automatica, 107*, 52–60.

Zhao, R., Zuo, Z., & Wang, Y. (2022b). Event-triggered control for networked switched systems with quantization. *IEEE Transactions on Systems, Man, and Cybernetics: Systems, 52*(10), 6120–6128.

Zhao, T., & Duan, G.-R. (2022). Fully actuated system approach to attitude control of flexible spacecraft with nonlinear time-varying inertia. *Science China Information Sciences, 65*(11), 212201.

Zhao, Y., Zhang, H., Chen, Z., Wang, H., & Zhao, X. (2022c). Adaptive neural decentralised control for switched interconnected nonlinear systems with backlash-like hysteresis and output constraints. *International Journal of Systems Science, 53*(7), 1545–1561.

Zhao, Y.-Z., Ma, D., & Zhang, Y.-W. (2023c). Adaptive asymptotic stabilization of uncertain nonstrict feedback nonlinear HOFA systems with time delays. *Nonlinear Dynamics, 111*(15), 14139–14153.

Zhao, Z., Liu, Y., Cai, S., Li, Z., Wang, Y., Hong, K.-S., and Li, H.-X. (2023d). Adaptive quantized control of flexible manipulators subject to unknown dead zones. *IEEE Transactions on Systems, Man, and Cybernetics: Systems, 53*(10), 6438–6447.

Zhao, Z., Tan, Z., Liu, Z., Efe, M. O., & Ahn, C. K. (2023e). Adaptive inverse compensation fault-tolerant control for a flexible manipulator with unknown dead-zone and actuator faults. *IEEE Transactions on Industrial Electronics, 70*(12), 12698–12707.

Zheng, S., & Li, W. (2019). Fuzzy finite time control for switched systems via adding a barrier power integrator. *IEEE Transactions on Cybernetics, 49*(7), 2693–2706.

Zhou, B. (2020a). Finite-time stability analysis and stabilization by bounded linear time-varying feedback. *Automatica, 121*, 109191.

Zhou, B. (2020b). Finite-time stabilization of linear systems by bounded linear time-varying feedback. *Automatica, 113*, 108760.

Zhou, B., & Shi, Y. (2021). Prescribed-time stabilization of a class of nonlinear systems by linear time-varying feedback. *IEEE Transactions on Automatic Control, 66*(12), 6123–6130.

Zhou, C., Wang, Y., Lv, M., & Wang, N. (2023). Neural-adaptive specified-time constrained consensus tracking control of high-order nonlinear multi-agent systems with unknown control directions and actuator faults. *Neurocomputing, 538*, 126168.

Zhou, J., Wen, C., Wang, W., & Yang, F. (2019). Adaptive backstepping control of nonlinear uncertain systems with quantized states. *IEEE Transactions on Automatic Control, 64*(11), 4756–4763.

Zhou, T., Liu, C., Liu, X., Wang, H., & Zhou, Y. (2021). Finite-time prescribed performance adaptive fuzzy control for unknown nonlinear systems. *Fuzzy Sets and Systems, 402*, 16–34. Theme: Control Engineering.

Zhu, W., & Jiang, Z.-P. (2014). Event-based leader-following consensus of multi-agent systems with input time delay. *IEEE Transactions on Automatic Control, 60*(5), 1362–1367.

Zhu, Z., Pan, Y., Zhou, Q., & Lu, C. (2021). Event-triggered adaptive fuzzy control for stochastic nonlinear systems with unmeasured states and unknown backlash-like hysteresis. *IEEE Transactions on Fuzzy Systems, 29*(5), 1273–1283.

Zou, A.-M., Liu, Y., Hou, Z.-G., & Hu, Z. (2023). Practical predefined-time output-feedback consensus tracking control for multiagent systems. *IEEE Transactions on Cybernetics, 53*(8), 5311–5322.

Zuo, Z., & Tie, L. (2014). A new class of finite-time nonlinear consensus protocols for multi-agent systems. *International Journal of Control, 87*(2), 363–370.

Zuo, Z., & Tie, L. (2016). Distributed robust finite-time nonlinear consensus protocols for multi-agent systems. *International Journal of Systems Science, 47*(6), 1366–1375.